# Gigantic Challenges, Nano Solutions

# Gigantic Challenges, Nano Solutions

## The Science and Engineering of Nanoscale Systems

**Maher S. Amer**

JENNY STANFORD
PUBLISHING

*Published by*

Jenny Stanford Publishing Pte. Ltd.
Level 34, Centennial Tower
3 Temasek Avenue
Singapore 039190

Email: editorial@jennystanford.com
Web: www.jennystanford.com

**British Library Cataloguing-in-Publication Data**
A catalogue record for this book is available from the British Library.

**Gigantic Challenges, Nano Solutions: The Science
and Engineering of Nanoscale Systems**

ISBN  978-981-4877-74-9 (Hardcover)
ISBN  978-1-003-14704-6 (eBook)

# Contents

# Preface

Two fundamental discoveries have recently started a new era of scientific research, the discovery of fullerenes and the development of single-molecule imaging capabilities. The discovery of fullerenes with their unique properties, highly versatile nature, and many potential applications in materials science, chemistry, physics, opto-electronics, biology, and medicine, has launched a new branch of interdisciplinary research known as "nanotechnology." This technology revolutionized the multibillion-dollar field of opto-electronics and is a key to wireless communications, remote sensing, and medical diagnostics and still has a lot to offer. The development of single-molecule imaging and investigating capabilities provided the means for studying reactions of complex material systems and biological molecules in natural systems.

The real importance of these discoveries is that they, synergized together, put forward the platform for what can be called "the next industrial revolution" in human history, "nanotechnology." Just as the quantum mechanics work of the 1930s led to the electronic material revolution in the 1980s, and as the fundamental work in molecular biology in the 1950s gave rise to the current biotechnology, it is believed that the emerging work in nanotechnology has the potential to fundamentally change the way people live within the next two decades. The ability to manipulate matter on the atomic level and to manufacture devices from the molecular level up will definitely have major implications. Among the advances and benefits foreseen for nanotechnology implementation are inexpensive energy generation, highly efficient manufacturing, environmentally benign materials, universal clean water supplies, atomically engineered crops resulting in greater agricultural productivity, radically improved medicines, unprecedented medical treatments and organ replacement, greater information storage and communication capacities, and increased human performance through convergent technologies. This means that nanotechnology is expected to revolutionize manufacturing and energy production, in addition to healthcare, communications, utilities, and definitely defense. Hence, nanotechnology will transform labor and the workplace, medical system, transportation, and power infrastructures. In short, nanotechnology will transform life, as we currently perceive it, and hence will have immense impact on human society.

In this book, we discuss nanotechnology and its attributes based on the observation that it represents a domain in which conventional materials perform in an unconventional way. We explain the phenomenon—the nanophenomenon—based on our current state of knowledge and try to predict its potentials and challenges.

Nanotechnology is based on certain building blocks that include fullerenes (in spheroidal, cylindrical, and sheet forms), nanocrystals, and nanowires and on characterization and imaging techniques capable of interrogating such building blocks. The book focuses on the fundamentals of nanoscale science and engineering. The main purpose is to show the reader that nanoscale systems are not necessarily small in size but can be huge. It is the thermodynamics of a system that dictates its behavior as a bulk (classic) or a small (nanostructured) system. The subject of this book is treated in eleven chapters. In Chapter 1, we introduce and discuss nanotechnology. Since social studies indicate that the majority of the public is not aware of the nature of nanotechnology, we felt obligated to start the book with a "layperson" level of introduction to nanotechnology. In the chapter, we show that as far as nature is concerned, nanotechnology is over 3 billion years old, and as far as humankind is concerned, nanotechnology was practiced several thousand years ago. On a more scientific level of discussion, in Chapter 2, we discuss the nanophenomenon. In this chapter, we introduce physical phenomena observed in a nanostructured system that are interesting as well as different from the well-known behavior of such systems in the classic (bulk) state. In Chapter 3, we explore nanotechnology and define nanodomain as the domain in which a system becomes a thermodynamically small system that can no longer be treated by classical thermodynamic laws originally observed and developed for bulk or classical systems. We show that once the size of the system is on the order of certain length-scales, the system becomes thermodynamically inhomogeneous and its thermodynamic potentials and functions become indefinable. In Chapter 4, we discuss such length-scales and show that materials behavior at such length-scales is unconventional and represents what can be termed as the "nanobehavior." We explain in this chapter how nanobehavior is related to the system's size as compared to thermodynamic inhomogeneity and not just to the system's physical dimensions. We also emphasize that unlike conventional bulk systems, nanosystems are very sensitive to perturbation effects. We provide evidence and show examples of the fact that significant changes in the behavior of a nanosystem can be observed as a result of minute perturbation fields affecting the system.

Since by now the reader should realize that nanostructured systems are mainly entropy- controlled systems, we have devoted Chapter 5 to discuss entropic effects and depletion forces as the major forces affecting the behavior of nanostructured systems. Since entropy as a great concept is typically related to system symmetry, we have devoted Chapter 6 to discuss symmetry and symmetry operations in one-, two-, and three-dimensional systems.

Chapter 7 is dedicated to fullerenes as the most promising building blocks for nanoscale systems. In this chapter, we present the history of fullerenes, their original predictions, and their initial discovery. We classify fullerenes based on their dimensionality into three classes: zero-, one-, and two-dimensional fullerenes. For each class we have devoted a chapter (Chapters 8–10), and we discuss the structure as well as the production and purification methods in each of them. For zero-dimensional fullerenes, we discuss $C_{60}$, $C_{70}$, and larger, or giant fullerenes. For one-dimensional fullerenes, we discuss single-, double , and multi-walled carbon nanotubes. For two-dimensional fullerenes, we discuss "graphene" in single- and multi-sheet forms. Nanofilms based on each building block and their properties are also introduced to the reader and discussed.

Finally, we dedicate Chapter 11 to discuss overview, potentials, challenges, and ethical consideration of nanoscale science and engineering, a subject that every scientist and engineer working in the field must pay attention to. As we mentioned in the beginning, nanoscale science and technology are destined to impact our way of living and it is crucial to understand the challenges and societal effects associated with it.

**Maher S. Amer**
Dayton, Ohio
July 2021

# Chapter 1

# Introduction and Overview

This chapter introduces nanotechnology and emphasizes the fact that it is more related to the thermodynamic behavior of small systems than to the physical dimensions of a system. It discusses the importance of entropic forces in such systems and presents examples of how such forces can alter the behavior of materials systems and enable them to exhibit unusual chemical, physical, electrical, optical, and mechanical properties. It identifies and discusses the building blocks of nanotechnology. It also gives examples of biological and natural utilization of nanostructured system as well as recent engineering applications of such systems.

## 1.1 Origins of Nanotechnology

Almost fifty years ago, on December 29, 1959, Richard P. Feynman,[1] a great physicist and, later, a Nobel Laureate, gave a lecture at the annual meeting of the American Physical Society at California Institute of Technology entitled "There's Plenty of Room at the

---

[1]**Richard P. Feynman** (1918–1988) was born in New York City on 11th May 1918. He studied at the Massachusetts Institute of Technology where he obtained his B.Sc. in 1939 and at Princeton University where he obtained his Ph.D. in 1942. He was a research assistant at Princeton University (1940–1941), professor of theoretical physics at Cornell University (1945–1950), visiting professor and thereafter appointed professor of theoretical physics at the California Institute of Technology (1950–1959). Feynman received the Nobel Prize in physics in 1965.

---

*Gigantic Challenges, Nano Solutions: The Science and Engineering of Nanoscale Systems*
Maher S. Amer
Copyright © 2022 Jenny Stanford Publishing Pte. Ltd.
ISBN 978-981-4877-74-9 (Hardcover), 978-1-003-14704-6 (eBook)
www.jennystanford.com

Bottom: An Invitation to Enter a New Field of Physics." The lecture was published later in *Engineering & Science*, Volume 23, No. 5, (February 1960) and was republished again in 1992 as the topic it first introduced overwhelmingly dragged the attention of many of the scientists, politicians, and the public across the globe.

The term "nanotechnology" was never used in Feynman's lecture; instead, Feynman spoke about *miniaturization* emphasizing the important scientific and economic aspects of our ability to make things *small*. Small machines those are capable of making even smaller ones. In his own words "although it is a very wild idea, it would be interesting in surgery if you could swallow the surgeon." Not to be misunderstood, Feynman emphasized that such a vision necessitates an ability to manipulate materials systems on a *small* scale. Feynman would not have missed the obvious and logical fact that atoms and molecules behave differently when arranged in a *small* system compared to their behavior in *large* or *bulk* systems. In his famous lecture Feynman said, "I can hardly doubt that when we have some control of the arrangement of things on a small scale we will get an enormously greater range of possible properties that substances can have." The obvious reason for that was "...Atoms on a small scale behave like nothing on a large scale, for they satisfy the laws of quantum mechanics. So, as we go down and fiddle around with the atoms down there, we are working with different laws, and we can expect to do different things." Hence, while Feynman did not explicitly spoke about what is referred to nowadays as "nanotechnology," he pointed out a new and important domain of physics where matter is investigated on a *new* scale at which quantum effects are dominating.

The term *nanotechnology* was actually coined in 1974 by Norio Taniguchi (1912–1988), a professor at the Tokyo Science University, Japan. The term *nano* is Greek for "dwarf." Professor Taniguchi's main interest was in high-precision machining of hard and brittle materials. He pioneered the application of energy beam techniques, including electron beam, lasers, and ion beams, to ultra-precision processing of materials. In his famous paper entitled "On the Basic Concept of 'Nano-Technology'," Prof. Taniguchi defined the field as, "'Nano-technology' mainly consists of the processing of, separation, consolidation, and deformation of materials by one atom or by one molecule." Prof. Taniguchi was mainly using the term to describe possibilities in precision machining for electronic industry to enable smaller and smaller devices down to the *nanometer* length scale. The prefix *nano* is known in the metric scale system to represent a billionth or $10^{-9}$ of a unit. In 1974, Prof. Taniguchi was interested in precision machining down to the nanometer level, which requires an ability to manipulate materials on the atomic or molecular level.

In 1986, K. Eric Drexler reused and popularized the term "nanotechnology" in a much broader prospective describing a whole new manufacturing technology based on molecular machinery. The premise was that such molecular machinery does exist, by countless examples, in biological systems, and, hence, sophisticated, efficient, and optimized molecular machines can be produced. In a series of books, Drexler described a number of possible molecular machinery suitable for a very wide range of applications. He also described "profiles of the possible" as well as "dangers and hopes" associated with the nanotechnology. Drexler was awarded a PhD in 1991. His work, indeed, triggered and inspired what is currently referred to as the nanorevolution. He adapted the viewpoint that although nanotechnology can be initially implemented by resembling biological systems, ultimately it could be based on pure mechanical engineering principles rendering nanotechnology as a manufacturing technology based on the mechanical functionality of molecular size components. Such mechanical components, i.e., gears, bearings, motors, and structural members, would enable programmable assembly with atomic precision. Figure 1.1 shows the three scholars Richard Feynman, who first envisioned nanotechnology, Norio Taniguchi who coined the term "nanotechnology," and Eric Drexler who popularized the term in a new prospective.

(a)        (b)        (c)

**Figure 1.1** The three scholars Richard Feynman, who first envisioned nanotechnology, Norio Taniguchi who coined the term "Nanotechnology," and Eric Drexler who popularized the term in a new prospective.

It has to be pointed out, however, that the pure mechanical viewpoint shaping Drexler's proposed vision led to a long and heated debate between him and Richard Smalley. Richard Smalley, a professor of chemistry at Rice University who shared the Nobel Prize in chemistry, 1996, with Robert Curl, Jr. and Sir Harry Kroto for discovering the $C_{60}$ molecule (fullerene [60]), had very well-founded reservations on applying pure mechanical engineering principles to nanomachinery, and on the premise of mechanical functionality of molecules. The debate was indeed a significant controversy about nanotechnology's meaning and possibilities. Drexler, later, backed off of his position on the bases that his original ideas have been misunderstood. Unfortunately, the debate left a negative impression on public view of the technology and, to a large extent, deepened the wrong concept that nanotechnology is the technology to make tiny (bug-like) machines capable of replicating themselves, working miracles, but could run amok. Giving the effective role media usually play on public viewpoint and understanding of science, it was concluded recently that the media has contributed to bounding nanotechnology by representing the term as a technology that trades on ideas of wonder as well as risk.

A good example demonstrating the general misunderstanding of nanotechnology is the winner illustration of the 2002 "Visions of Science Award" by Coneyl Jay shown in Figure 1.2. The illustration shows how the public, in general, imagined what nanotechnology is all about; a bug-like tiny machine injecting *stuff* in a red blood cell! Unfortunately, that was the general impression on nanomedicine.

Such a simplistic understanding of nanotechnology bred enormous public concerns and suspicion. Well founded and justified concerns regarding the impact of nanotechnology on scientific, economic, ethical, and societal aspects of humankind future were raised and are still being debated. We will discuss these important issues in later sections of this chapter. At this point, however, it is beyond any doubt that what we decided to call a *dwarf (nano)* turned out to be a *giant*.

**Figure 1.2** Science art illustrating the perception of nanobots. Copyright © Coneyl Jay/Science Photo Library.

## 1.2 What Is Nanotechnology?

Nanotechnology has been described as the third industrial revolution in human history. As each of the previous industrial revolutions, it is expected to have a huge and long-term impact on all aspects of human life. In addition, and not surprisingly, the new technology has not been very well understood in some cases and has been misunderstood in many other cases. According to the USA National Science Foundation 2007 released statistics, the majority of Americans (54%) have heard "nothing at all" about nanotechnology. In this section, we will address the nature of nanotechnology in very simple, even layperson, terms. As Albert Einstein pointed out, one can claim knowledge of a subject only when one is capable of explaining the subject to one's grandmother. Time has changed since the Einstein era and many of nowadays grandmothers have advanced degrees. While this makes it easier for new generations to claim knowledge, it might be the time to change the rule of knowledge claiming to state that one may claim knowledge of a subject only if one is capable of *correctly* explaining the subject to the public.

### 1.2.1 What Can Nanotechnology Do for Us?

Everything we deal with is either human-made (made by humans) or natural (made by nature). For example, a car is a man-made transportation means, while a horse is a natural *one*. Currently, cars are faster and much stronger than horses. However, cars are still not capable of sensing the danger down the road as horses do. Also, cars cannot take their passenger home while the passenger is asleep as horses do. In addition, horses are much safer to travel by since none of us has ever heard about an accident between two horses resulting in rider's life loss! To this end, we can describe nanotechnology as a new level of knowledge that could enable us to bridge the gap between the capabilities of man-made and natural things. This would result in a new generation of regular size, and not tiny, cars capable of sensing the danger, driving home, and reducing or eliminating accidents due to operator errors. In addition, a more important and a major difference between man-made and natural things is their efficiency. Over billions of years, nature mastered the art of efficient design and operation. Humans, however, are still at the beginning of a learning curve in those regards. For example, the best gasoline car engine we currently make has an efficiency of 25% ∼ 30%. Mechanical efficiency of athletes during running was measured to range between 47% and 62%. Other species and natural processes can even reach higher efficiency than that. This clearly demonstrates how crucial nanotechnology can be considering the energy crises our civilization is currently facing.

### 1.2.2 Where the Name "Nano" Came From?

The word *nano* is Greek for *dwarf*. This word was actually used to indicate the length unit equal to one billionth of a meter ($10^{-9}$ meter). In order to have a good idea of what this length actually is, let us consider a typical single human hair. This is about 50 to 100 micron which is 50 to 100 millionth of a meter. Hence, a single human hair would be 50 to 100 thousand times larger than a nanometer. Atoms and molecules are typically measured by a unit called the Angstrom, which is one tenth of a nanometer, or one ten-billionth of a meter. Fifty years ago, Feynman predicted, and more recently, many scientists observed that the behavior of material clusters on the 1 ~ 100 nanometer scale is essentially different from that of larger clusters that we currently use. It is scientifically sound to say that on the nanometer level, the laws of nature controlling materials behavior are different, hence new phenomena can be observed. This is where the name "nanotechnology" came from.

### 1.2.3 Does Every Nanosystem Have to Be So Small?

The answer for this question is absolutely not. In fact, most of biological systems, including ourselves, are nanosystems. This is in the sense that these systems on a certain level operate according to nanophysics and nanochemistry laws. In fact, humans in their daily life activities obey two different sets of laws; traditional laws of physics that we already know, and nanoscale laws that we are still exploring. For example, if one jumps up, one's body will follow the gravitational law that was first identified by Isaac Newton in the fifteenth century, and one will, according to this law, fall back down. However, as one breathes one's blood exchanges carbon dioxide for oxygen at the lungs and does the opposite at the cells according to a different set of laws that we refer to here as nanolaws. The blood component in charge of the exchange (known as hemoglobin) is much larger than a nanometer, and so are our body cells, and definitely we and all other breathing creatures are. A nanosystem, or more appropriately, a nanostructured system is a system that is made of components that operate according to nanolaws regardless of the system size. Good examples to illustrate this point are butterfly wings and opal stones shown in Figure 1.3. The beautiful colors of butterfly wings and opals are due to light reflection by nanostructures and are not due to pigmentation. Different colors are due to different nanostructures. The opal example tells us that nanostructures are not limited to biological systems. In fact, over billions of years, nature mastered nanomanufacturing techniques as the best and most efficient techniques to build sophisticated and efficient products.

**Figure 1.3** Nanostructures responsible for the amazing colors in butterfly wings and opal.

## 1.2.4 The Properties of Matter Change by Entering the Nanodomain

A good example to illustrate how the properties of matter change as it enters the nanodomain is water. In its bulk form, water is a liquid that every one of us is familiar with. It is a colorless odorless liquid that is heavier than air. Hence, gravitational laws are in control of the system and it fills our lakes, seas, and oceans. As water evaporates, due to heat effects, in the form of single and tiny clusters of molecules, gravitational laws are no longer in charge. The bulk laws are actually overruled by the new *nanodomain* laws. Water in the form of tiny clusters becomes airborne. These tiny clusters of water can accumulate and form huge clouds containing enormous amounts of water but still can be transported by wind over very long distances. Controlled by weather conditions, the tiny clusters of water in the clouds can grow into bigger and bigger clusters until they depart the nanodomain and enter the bulk domain again in the form of water droplets. Once in the bulk domain, gravitational laws will take control again, and water droplets will fall as rain. To this end, it is very clear that nature has utilized nanotechnology, for billions of years, in transporting enormous amounts of water over great distances very efficiently. It is interesting to note that even with our current *advanced* technologies, as we like to call it, we are not capable of carrying on such a transportation operation as efficient, if at all. It might be wisely and timely to learn from nature.

## 1.2.5 Has Nanotechnology Been Used Before?

While nature has been utilizing nanotechnology in building biological systems, as we mentioned before, for almost 3.7 billion years, humans have also used nanotechnology before. The mysterious optical behavior of the famous Lycurgus Cup (400 AD, Figure 1.4) is a good example of nanotechnology effects on optical properties of matter. This Roman cup is made of ruby glass. When viewed in reflected light, for example in daylight, it appears green. However, when a light is shone into the cup and transmitted through the glass, it appears red! Recently this mysterious behavior was investigated and it was found that while the chemical composition of the glass of Lycurgus cup is almost the same as that of modern glass, the fascinating optical behavior is totally due to gold nanoparticles within the cup glass.

**Figure 1.4** Lycurgus Cup (British Museum; AD fourth century). This Roman cup is made of ruby glass. When viewed in reflected light, for example in daylight, it appears green. However, when a light is shone into the cup and transmitted through the glass, it appears red. Reproduced with kind permission from Leonhardt, *Nat. Photonics*, 2007, 1, 207. Copyright Nature Publishing Group, 2007.

In addition, it was revealed recently that the famous Damascus saber (14th to 16th century era), with its traditional wavy patterns on the surface (see Figure 1.5), owes its superior strength and performance (as the first self-sharpening blades) to the presence of carbon nanostructures in the form of tubes and wires within its alloy.

**Figure 1.5** Typical wavy patterns on the surface of a Damascus saber. The structure was found to be due to carbon nanotubes in the saber alloy. Reproduced with kind permission from Levin et al., *Cryst. Res. Technol.*, 2005, 40, 905. Copyright Wiley-VCH, 2005.

Nanotechnology, namely nanomedicine, was used even much earlier in human civilization, most probably unintentionally. Finely ground gold particles in the size range 10 to 500 nanometers can be suspended in water. Such suspensions were used for medical purposes in ancient Egypt over five thousand years ago. In Alexandria, Egyptian alchemists used fine gold particles to produce a colloidal elixir known as "liquid gold" that was intended to restore youth!

In 1856, Michael Faraday (1791–1867) independently prepared colloidal gold which he called "divided state of gold." Faraday's samples are still preserved in the royal institution. Also, in 1890, the work of the German bacteriologist Robert Koch (1843–1910) showed that gold compounds inhibit the growth of bacteria. He was awarded the Nobel Prize for medicine in 1905. Figure 1.6 shows the two scholars who pioneered the preparation and medical applications of gold nanoparticles.

(a)                              (b)

**Figure 1.6** The two scholars pioneered gold nanoparticles and their medical applications (a) Michael Faraday and (b) Robert Koch. Photos courtesy of the Nobel Foundation.

## 1.2.6 Could Nanoscale Engineering Be Developed Earlier?

The development and progress of any branch of science is a matter of probability. Three main ingredients have to exist; a vision or a theory, characterization and diagnostic techniques, and a material system, or in other words, building blocks for experimental verification. In addition, economic, societal, and political environments have to be suitable to enable the synergy of the three main ingredients into one successful and sustainable endeavor. If we consider nanotechnology, Feynman's vision was born as a natural consequence of the successful theoretical physics developments in quantum mechanics during the 1930s. The vision was most probably delayed by a decade because of World War II. Feynman's vision, however, had to wait till the 1980s for the discovery of the scanning tunneling microscope in 1981 and the fullerene building blocks in 1985, for the triad to be complete. On one hand, scanning tunneling microscopy was the tool that allowed the required resolution and the manipulation capability of single atoms and molecules. On the other hand, fullerene discovery opened the door for the production and investigation of suitable material building blocks. Figure 1.7 shows the two scholars and Nobel Laureates Gerd Binnig and Heinrich Rohrer who invented the scanning tunneling microscope (STM), and Figure 1.8 shows the three scholars and Nobel Laureates Curl, Smalley, and Kroto who first discovered the fullerene molecules.

(a)  (b)

**Figure 1.7** The two scholars and Nobel Laureates Gerd Binnig and Heinrich Rohrer who invented the scanning tunneling microscope (STM). Photos courtesy of the Nobel Foundation.

(a)  (b)  (c)

**Figure 1.8** The three scholars and Nobel Laureates (a) Curl, (b) Smalley, and (c) Kroto, who first discovered and isolated $C_{60}$ fullerene molecules. Photos courtesy of the Nobel Foundation.

Before we proceed to our scientific discussions and equations, a crucial point has to be made very clear. While nanoscale science and engineering will definitely play a major role in the future of human civilization, no one should suffer the illusion that it will provide solutions for all human-kind problems. It is unfortunate that some people in the scientific field, for different reasons, portrayed nanotechnology as capable of making every human on planet earth younger, more beautiful, and richer than what they currently are! In fact, as much as nanotechnology will provide solutions for our challenges, it will also impose new challenges on us. Many of such challenges cannot be even predicted at the present time.

## Problems

1. Who was the first scholar to drag attention to the different behavior of matter in small systems?
2. When and who coined the term "nanotechnology"?
3. What was the original field for which the term "nanotechnology" was coined for?
4. Is it possible to apply pure mechanical functionality at a molecular level?
5. Would a molecular machine perform as expected?
6. What is nanotechnology as you understand it?
7. What can nanotechnology enable us to achieve?
8. Does every nanosystem have to be small in size?
9. Has nanotechnology been used before our current time? Give examples.
10. Having the idea of small systems since the 1960s, why nanotechnology could not be developed then?

# Chapter 2

# Nanophenomena

Several phenomena have been recently observed that can be attributed to the behavior of matter in the nanodomain. In this chapter we will mention some of such very interesting phenomena and their immediate impact on the development of our technology.

## 2.1  Optical Phenomena

We mentioned before how nature utilized nanostructure in butterflies' wings and opal stones to develop amazing colors. Recently, gold nanoparticle solutions were shown to yield different colors depending upon the gold particle size and shape. Figure 2.1 shows solutions of nanogold particles of different sizes in water. It is interesting to realize that for gold particles size of 2 nm (far left) the particle size is too small to affect the light interaction with the solution resulting in a solution with optical properties similar to those of water. As the gold particles increase in size, their interaction with the light becomes significant causing the observed different colors of the solution. It is widely known that light interaction with the solutions in this case is strongly affected by the electronic structure of the solution. Figure 2.1 is a good example of particle size effect on its electronic structure as reflected by its interaction with light.

*Gigantic Challenges, Nano Solutions: The Science and Engineering of Nanoscale Systems*
Maher S. Amer
Copyright © 2022 Jenny Stanford Publishing Pte. Ltd.
ISBN 978-981-4877-74-9 (Hardcover), 978-1-003-14704-6 (eBook)
www.jennystanford.com

**Figure 2.1** Monodispersed gold nanoparticles of different sizes (2 nm – 200 nm) in water. Note the different colors due to different light interactions dominated by particle size. Courtesy of Ted Pella, Inc. available on the web at http://www.tedpella.com/gold_html/goldsols.htm.

Such an optical nanophenomenon is not limited to metallic particles. Semiconducting nanoparticles (sometimes referred to as quantum dots) also show very interesting dependence of their electronic structure on the particle size as reflected by their interaction with visible light and electromagnetic radiation in general. Figure 2.2 shows cadmium selenide (CdSe) nanoparticles of different sizes in solution observed under radiation with visible light (bottom) as well as dark light (ultraviolet radiation) (top). It is clear from the photo that the size of the quantum dot affects its electronic structure as reflected by the different solution colors. More interestingly, as shown in the top photo, irradiation with ultraviolet radiation (dark light) cause the solution to irradiate different colors in the visible range. This is another interesting example of an optical nanophenomenon with potential applications in harvesting the ultraviolet portion of solar energy.

It was also interesting to realize that not only particle size but also particle shape affects its optical properties and controls the system absorption behavior. It was found that gold nanorods interact differently with light based upon the rod length to diameter (aspect) ratio. Depending upon their aspect ratio, gold nanorods were found to absorb light with a maximum located at different wavelengths in the visible and the near infrared spectra range. Figure 2.3 shows the maximum absorption wavelength for gold nanorods of different aspect ratios. It is important to note that gold nanospheres (denoted by an aspect ratio 1 in Figure 2.3) absorb light at different wavelengths as well.

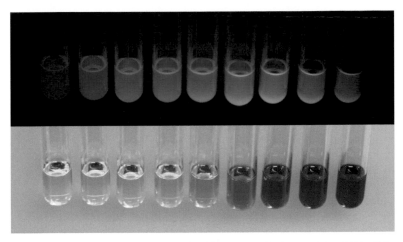

**Figure 2.2** CdSe quantum dots of different sized observed in visible light (bottom) as well as dark light (ultraviolet radiation) (top). Courtesy of the University of Wisconsin-Madison, Materials Research Science and Engineering Center. Available on the web at: http://mrsec.wisc.edu/Edetc/background/quantum_dots/index.html.

In addition, the maximum absorption wavelength was also shown to depend on the orientation of the gold nanorods in respect to the polarization direction of the light. The maximum wavelength of light absorption for light polarized parallel to the nanorods axis ( $\lambda_{max}^{\parallel}$ ) was found to be different from that for a light polarized in a direction normal ( $\lambda_{max}^{\perp}$ ) to the nanorods axis direction. The phenomenon was related to a resonance effect in the particle surface plasmons activity, the surface plasmon resonance (SPR) effect that, in turn, depends on the geometry of the nanorods was found to control the position of to a maximum in the optical adsorption spectrum of such nanoparticles. Figure 2.4 shows the dependence of the ( $\lambda_{max}^{\parallel}$ ) on the aspect ratio of the gold nanorods. It has been shown that ( $\lambda_{max}^{\parallel}$ ) strongly shifts linearly to longer wavelengths as the rod aspect ratio increases while ( $\lambda_{max}^{\perp}$ ) shifts down to lower wavelengths as the rod aspect ratio increases reaching a plateau for rods with aspect ratio higher than five.

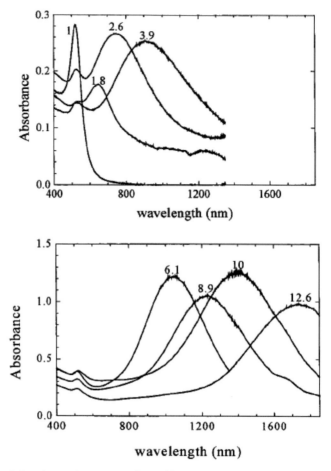

**Figure 2.3** Absorption spectra for gold nanorods of different aspect ratios. Note the shift in the maximum absorption wavelength at different aspect ratios denoted at the top of each curve. Aspect ratio 1 denotes gold nanospheres. Reproduced with kind permission from Van der Zande et al., *J. Phys. Chem. B*, 1999, 103, 5761. Copyright American Chemical Society, 1999.

Optical properties nanospheres with core shell structures also shows interesting optical nanophenomenon. Figure 2.5 shows the different types of nanoparticles with shell/core structures that has been produced and investigated. The types ranged from short molecules tethered to a surface modified core particle as shown in Figure 2.5a, to multi-shell (onion-like) type of structure as shown in Figure 2.5e.

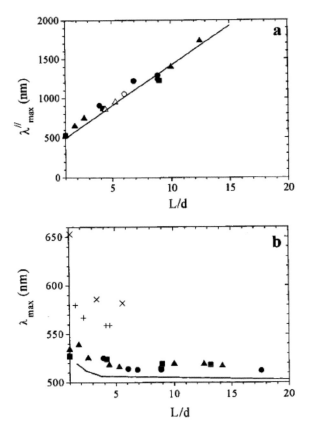

**Figure 2.4** Dependence of the plasmon absorption maxima for (a) longitudinal ($\lambda_{max}^{\parallel}$) and (b) transverse ($\lambda_{max}^{\perp}$) polarizations conditions on the gold nanorods aspect ratio for gold nanorods dispersed in water. Dark squares, circles, and triangles represent experimental data obtained from rods with diameters of 12, 16, and 20 nm, respectively. Open symbols indicate samples with narrow size distribution. + and × symbols represent rods with 86 nm and 120 nm diameters, respectively. Reproduced with kind permission from Van der Zande et al., *Langmuir*, 2000, 16, 451. Copyright American Chemical Society, 2000.

Different types of materials (i.e., metallic, dielectric, etc.) were also used to make such nanoparticles. Interestingly enough, the optical absorption of these nanoparticles showed dependence on the relative dimensions of the core and shell as well as the exact structure of the particle. It was reported that the observed optical resonance and maximum absorption wavelength can be varied over hundreds of nanometers in wavelength across the visible and

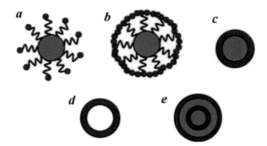

**Figure 2.5** Variety of core/shell particles produced and investigated; (a) surface decorated particles, (b) heavier surface decoration forming a shell around the core, (c) solid shell around a core, (d) quantum bubbles, and (e), multi-layered particles.

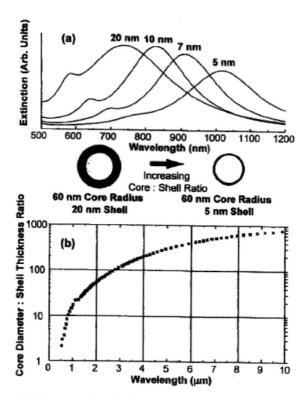

**Figure 2.6** (a) Theoretically calculated optical resonance of gold nanoshells of different thicknesses on a 60 nm silica core and (b) calculated optical resonance wavelength for the same samples in (a). Reproduced with kind permission from Oldenburg et al., *Chem. Phys. Lett.*, 1998, 288 (2), 243–247. Copyright Elsevier, 1998.

into the infrared region of the spectrum. Such results point out the crucial possibility of engineering such structured on the nanoscale in order to produce new class of matter with well-designed optical properties for a number of optoelectric applications. Figure 2.6a shows the theoretical calculations for the resonance optical absorption spectra of a 60 nm silica core with gold shell of different thicknesses. Figure 2.6b shows the calculated optical resonance wavelength for the same nanoparticles. It is intriguing to realize that the resonance wavelength increases (shifts from the visible into the infrared spectral region) as the shell thickness is reduced from 20 nm to 5 nm. Figure 2.7 shows the different colors obtained from gold nanoshell particles with different shell/core relative dimensions dispersed in water in support of the theoretical calculation results shown in Figure 2.6.

**Figure 2.7** A photograph showing colors of gold nanoshell particles with different shell thicknesses dispersed in water. Photo by C. Rodloff. Courtesy of Professor Halas's group of Rice University, USA.

In addition, self-assembled nanoparticles monolayers exhibited optical absorption spectra that not only depend of the properties of the individual particles but also on the interactions between these particles, as well as on their interaction with their environment. With their optical properties depending on the shape, size, composition, and their exact structure in addition to their interaction with each other and their environment, nanoparticles represent a very fertile ground and a virtually unexplored frontier for future investigations. Figure 2.8 (left) shows, a schematic drawing of a multi-layer film assembled using layer-by-layer technique of silica-coated gold nanoparticles embedded in cationic polyelectrolyte on a glass substrate (1 = glass substrate; 2 = cationic polyelectrolyte;

3 = nanoparticles), and Figure 2.8 (right) shows photographs of transmitted (top) and reflected (bottom) colors from the multi-layer thin films with varying silica shell thickness.

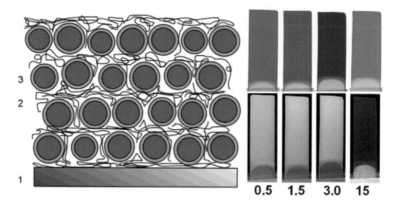

**Figure 2.8** Left: Schematic drawing of a multi-layer film assembled using layer-by-layer technique of silica-coated gold nanoparticles embedded in cationic polyelectrolyte on a glass substrate (1 = glass substrate; 2 = cationic polyelectrolyte; 3 = nanoparticles). Right: Photographs of transmitted (top) and reflected (bottom) colors from the multi-layer thin films with varying silica shell thickness. Reproduced with kind permission from L. Liz-Marzán, *Mater. Today,* 2004, 7, 26. Copyright Elsevier, 2004.

## 2.2   Electronic Phenomena

It is very well known that electronic and optical properties of matter are very well correlated. The previously mentioned unusual and versatile optical phenomena observed for nanoparticles are simply the result of an unusual and versatile electronic structure of such new class of matter, or of such old matter observed in a new domain. The electronic structure of metal cluster of small sizes has been an active and fertile field of investigation for over three decades. There has been good progress in understanding the nature of small metal clusters and a realization that they cannot be simply treated as minute elements of a block of a metal. This is based on research findings showing that the conduction band present in a bulk metal will be absent and instead there would be discrete states at the band edge. In other words, the electronic band structure of a metal was found to change significantly with the size of the metal cluster once in the nanodomain. Figure 2.9 illustrates the electronic band structure of metals in the bulk form and in the atomic form passing through the nanodomain. It is important to note that a metal single atom is, in fact, an insulator, in the nanodomain, the metal cluster has the electronic band structure of a semiconductor, and in the bulk, metals, as we know, are conductors with overlapping valence and conduction bands.

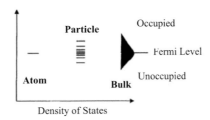

**Figure 2.9**   Electronic structure for a metal as an atom, a particle, and in the bulk state.

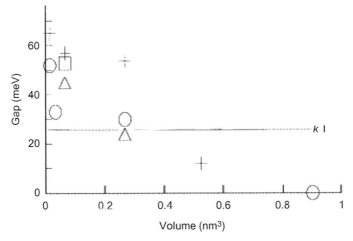

**Figure 2.10**   Conduction gap observed in small clusters of gold (Au), palladium (Pd), cadmium (Cd), and silver (Ag) as a function of cluster volume. Adapted with kind permission from Vinod et al., *Chem. Phys. Lett.*, 1998, 289, 329. Copyright Elsevier, 1998.

The very well-known conductive nature of bulk metals is not a given fact anymore as metals assume smaller and smaller clusters. Energy gaps in the electronic structure of conductor clusters with a value dependent of the cluster size become observable. Figure 2.10

shows the conduction gap observed in small clusters of gold (Au), palladium (Pd), cadmium (Cd), and silver (Ag) as a function of cluster volume. Energy band gap for all tested metals increased with the reduction of the metal cluster volume.

The dependence of electronic structure, as reflected in size of the energy band gap, on matter cluster size was not limited to metal clusters. In fact, similar phenomenon was observed for semiconductor clusters as well. The traditional band energy gap in semiconductors was found to be dependent of the cluster size as well. Figure 2.11 shows the energy band gap for different semiconductors. It can be inferred from the experimental results shown in Figure 2.11 that the energy band gap remains size independent till the cluster becomes smaller than a critical size below which the energy band gap increases sharply, hence, its electric conductivity decreases sharply as well. Such critical size value apparently depends on the value of the energy gap in the bulk state of the semiconductor. The smaller the bulk state energy band gap of the semiconductor is, the larger the critical size of the cluster. It is also clear from the results that the smaller the bulk state energy band gap of the semiconductor is, the more sensitive the size dependence becomes as the semiconductor cluster becomes smaller than its critical size.

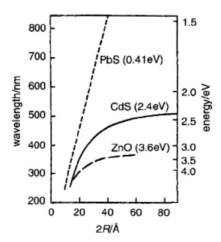

**Figure 2.11** Size-dependence of the wavelength of the absorption threshold in semiconductors. Values in parentheses represent the energy band gap of the semiconductor in the bulk state. Reproduced with kind permission from A. Henglein, *Chem. Rev.*, 2002, 89, 1861. Copyright VCH, 1995.

## 2.3   Thermal Phenomena

Recent theoretical studies showed that as a solid enters the nanodomain, its thermodynamic constants such as, Debye temperature, and specific heat capacity are no longer constant but become size dependents. These quantities were also shown to change with the system temperature and the nature of the chemical bonds involved. While surface effects and quantum confinement explanations are most appealing, the exact mechanisms of such intriguing phenomenon and the correlation among these quantities, however, are still to be clarified. The fact that matter on the nanometer scale can exist in one-dimensional forms, such as nanowires and nanotubes, adds an appealing dimension to thermal transport in this class of matter. For example, thermal conductivity of individual single crystalline intrinsic silicon nanowires with diameters of 22, 37, 56, and 115 nm was recently measured over a temperature range of 20–320 K, and was found to be more than two orders of magnitude lower than the bulk silicon value. Figure 2.12a shows the measured thermal conductivity for silicon nanowires of different diameters ranging between 22 nm and 115 nm. The dependence of the thermal conductivity of the nanowires on the wire diameter is very clear. Figure 2.12b shows the logarithm of thermal conductivity of the same nanowires in the low temperature range plotted against the logarithm of temperature. Interestingly enough, the data show that the behavior of the large diameter nanowires fits the well-known Debye $T^3$ law quite well in this temperature range. However, for the smaller diameter wires, 37 and 22 nm, the power exponent gets smaller as the diameter decreases and deviation from Debye $T^3$ law is clear.

Another interesting nanophenomenon is the observed effect of alumina nanoparticles on the thermal conductivity of fluids once suspended in them. Experimental data obtained for alumina nanoparticles of different shapes in a fluid consisting of equal volumes of ethylene glycol (EG) and water showed an enhanced thermal conductivity. Figure 2.13 shows the measured enhancement in fluid thermal conductivity as a function of the nanoparticles' concentration for different nanoparticles shapes of alumina. It is clear from the figure that the experimental results agree well with the predictions of recently developed theoretical models.

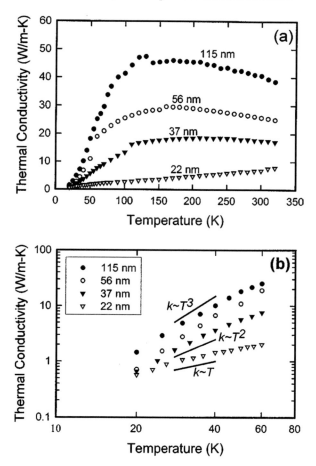

**Figure 2.12** (a) Measured thermal conductivity of different diameter Si nanowires. The number beside each curve denotes the corresponding wire diameter. (b) Low temperature experimental data on a logarithmic scale. Also shown are $T^3$, $T^2$, and $T^1$ curves for comparison. Reproduced with kind permission from Li et al., *Appl. Phys. Lett.*, 2003, 83, 2934. Copyright American Institute of Physics, 2003.

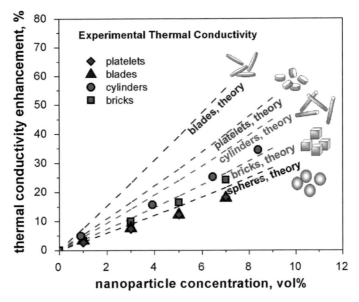

**Figure 2.13** Experimentally measured thermal conductivity of alumina nanofluids in EG/H2O compared to predictions of theoretical models. Reproduced with kind permission from Timofeeva et al., *J. Appl. Phys.*, 2009, 106, 014304. Copyright American Institute of Physics, 2009.

## 2.4   Mechanical Phenomena

Mechanical behavior of small systems is another anomalous phenomenon that attracted attention of many researchers and is still under investigation. For example, nanostructured metals showed an interesting deviation from the historically accepted Hall–Petch relationship expressed as;

$$\sigma_y = \sigma_o + kd^{-1/2}$$

where $\sigma_y$ is the yield strength, $\sigma_o$ is the friction stress, $k$ is a constant, and $d$ is the grain size. The deviation from Hall–Petch relationship was reported for the first time in 1989 for nanostructured copper, and was since referred to as the *Inverse Hall–Petch effect*. It has been reported for many other metals with grain size ranging between 1000 nm and 100 nm. Figure 2.14 shows experimental plots depicting the trend of yield stress with grain size for different metals as compared to the conventional Hall–Petch response: (a) copper, (b) iron, (c) nickel, and (d) titanium.

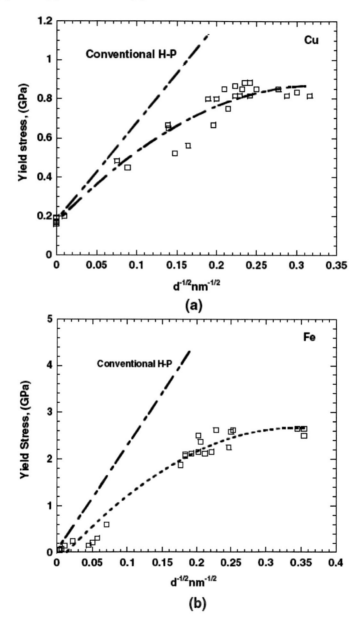

(a)

(b)

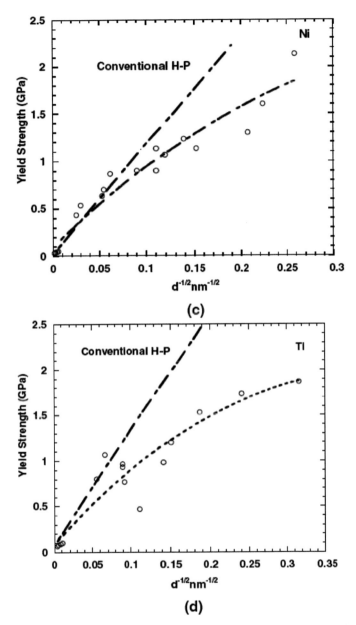

**Figure 2.14** Plots showing the trend of yield stress with grain size for different metals as compared to the conventional Hall–Petch response: (a) copper, (b) iron, (c) nickel, and (d) titanium. Reproduced with kind permission from Meyers et al., *Prog. Mater Sci.*, 2006, 51, 427. Copyright Elsevier, 2006.

In addition, while a reduction in grain size leads to an increase in ductility in conventional metals, ductility was found to be small for systems with grain size less than 25 nm. Other anomalous behavior was reported for nanostructured materials regarding their creep, fatigue, as well as their strain rate sensitivity.

## Problems

1. State three optical nanophenomena.
2. What are the factors that affect the optical properties of a nanosystem?
3. How does the energy ban gap of a metal cluster depend on the cluster volume?
4. Which was found to have a higher value, the thermal conductivity of bulk silicon or that of silicon nanowires?
5. How does the thermal conductivity of nanowires depend on wire diameter?
6. State the effect of adding alumina nanoparticles on thermal conductivity of EG–water mixtures.
7. What is the Hall–Petch relationship? What class of materials does it apply for?
8. Was the Hall–Petch relationship found to apply for nanosized grain size metals?

# Chapter 3

# Bulk Systems and Nanoscale Systems

Fifty years after Feynman's announcement of his vision, now we know that it is not merely obeying the quantum mechanical laws that makes materials behave differently in small and large systems but also the system size dependence of the nature and relative importance of forces controlling the system and dictating its behavior. We now know that as the system size becomes smaller and smaller, gravitational forces, a major player in large systems, start to lose their control on the system and other forces, such as surface tension, van der Waal, and entropic or depletion forces start to take control. Example of this is smoke. Smoke is defined as a mixture of "small" solid particles and a gas. Every time you watch smoke rising up, the solid particles in that smoke are simply defying gravitational forces since their behavior is controlled by another set of forces as we will discuss latter. The terms *small* and *large* or *nanoscale* and *bulk* have been mentioned frequently so far. It might be a good idea to start our discussion of nanoscale science and engineering by defining and differentiating between these two important terms.

## 3.1 Thermodynamics of Large and Small Systems

It is well known that thermodynamics and statistical mechanics, our principal theoretical tools for understanding the physics and behavior of material systems, are mainly based on the assumption

*Gigantic Challenges, Nano Solutions: The Science and Engineering of Nanoscale Systems*
Maher S. Amer
Copyright © 2022 Jenny Stanford Publishing Pte. Ltd.
ISBN 978-981-4877-74-9 (Hardcover), 978-1-003-14704-6 (eBook)
www.jennystanford.com

that the system under consideration is infinitely large. It is important to note, here, that "large" does not refer to the system physical size but to the notion that the system is essentially uniform at equilibrium even if it has multiple phases. Uniformity in this context is the uniformity of thermodynamic functions in the system. Thermodynamic functions of common interest include pressure ($p$), temperature ($T$), chemical potential ($\mu$), internal energy ($U$), and free energy ($F$). Hence, in classical thermodynamic, a system at equilibrium must has uniform pressure ($p$), temperature ($T$), and chemical potential ($\mu$) everywhere within the system, hence its internal energy ($U$), and free energy ($F$) are constants.

If the uniformity, of thermodynamic functions, condition is not met, then, regardless of the actual physical dimension of the system or its number of molecules, the system can no longer be considered a *large* (bulk, or classical) system and has to be treated as a *small* (nanoscale) system for which *classical* definitions of the aforementioned thermodynamic functions are no longer valid.

For small thermodynamic systems, the *quasi-thermodynamic assumption* or sometimes called the *point thermodynamic approximation* states that it is possible to define unique and useful thermodynamic functions at a point ($r$). This leads to the definition of *local thermodynamic functions* at any point ($r$) in a system as; *local pressure $p(r)$*, *local temperature $T(r)$*, and *local chemical potential $\mu(r)$*. In addition, three local densities can be defined as *local number density, $\rho(r)$*; *local internal energy density, $\phi(r)$*; and *local free energy density, $\psi(r)$*.

It is important to note that the local thermodynamic functions and densities can be defined by describing, around the point ($r$), a small volume ($\delta V$) that contains a number of molecules ($\delta N$), and has an internal energy ($\delta U$), and free energy ($\delta F$) as follows:

Local temperature $T(r)$ can be defined by knowing the translation kinetic energy ($\delta K$) of molecules within the small volume ($\delta V$) as

$$\frac{3k}{2}T(r) = \lim_{\delta V \to 0}\left(\frac{\langle \delta K \rangle}{\langle \delta V \rangle}\right) \tag{3.1}$$

Local chemical potential can be determined as follows:

$$\frac{\mu(r)}{kT} = \ln\left[\Lambda\rho(r)\right] - \ln\langle\exp\left[-\frac{u(r)}{kT}\right]\rangle \tag{3.2}$$

where $\Lambda$ is the de Broglie wavelength. While the de Broglie wavelength is not essential in the current discussion, it is good to know that it is the equivalent wavelength of a moving particle and can be calculated as

$$\Lambda = \frac{h}{mv} \tag{3.3}$$

Here, $h$ is Planck's constant ($6.626 \times 10^{-34}$ J·s), $m$ is the moving particle mass (kg), and $v$ is its speed (m/s).

Local number density can be determined as

$$\rho(r) = \lim_{\delta V \to 0} \left( \frac{\delta N}{\delta V} \right) \tag{3.4}$$

Local internal energy density can be determined as

$$\phi(r) = \lim_{\delta V \to 0} \left( \frac{\delta U}{\delta V} \right) \tag{3.5}$$

Local free energy density can be determined as

$$\psi(r) = \lim_{\delta V \to 0} \left( \frac{\delta F}{\delta V} \right) \tag{3.6}$$

Recent studies raised questions regarding the ability to measure local temperature on the nanoscale. With recent advances enabling temperature measurement on the nanoscale with a "nanothermometer," theoretical studies addressed the fundamental question of how meaningfully temperature can be defined on the nanoscale. Special attention was given to the smallest system size for which local temperature can be said to exist and can be defined. Fundamental issues regarding the definition of local temperature on such a small scale, and the agreement between theory and experiments are still being investigated and to be resolved.

To this end, we know that a thermodynamically *homogeneous* system is large and its physics can be described by classical thermodynamics and statistical mechanics. Equations of state can be developed for such a system; hence, its physical behavior can be predicted. A thermodynamically *inhomogeneous* system, however, has to be treated as a small system and local thermodynamic functions has to be determined for such a system in order to enable accurate description of its physics. *This raises a question regarding how determinable local thermodynamic functions are and the scale limit, if any, at which such functions become accurately determinable. To elucidate the idea, let's consider the following example.*

**Example 3.1:** Consider the local number density $\rho(r)$ function defined in Eq. 2.3. Figure 3.1a shows a system that contains one mole of $CO_2$ gas contained in a volume of 1 m³. In such a case, the number density of the system is clearly 1 mole/m³. Now, let's refer to Figure 3.1b. Here the system considered is a cube containing one molecule of $CO_2$ with a side length of 3.52 Å which is the exact length of the $CO_2$ linear molecule as shown in Figure 3.1b. In this case, the number density $\rho(r)$ can be calculated as

$$\rho(r) = (N/V) = 1/(3.52 \times 10^{-10})^3 = 2.29 \times 10^{28}\,\text{molecule/m}^3$$
$$= 3.8 \times 10^4\,\text{mole/m}^3.$$

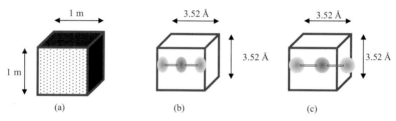

Figure 3.1   (a) A system containing one mole, (b) a system containing one molecule, and (c) the system in (b) as the molecule vibrates.

However, as the $CO_2$ molecule vibrates and stretches, the number density in our defined system (the cube with 3.52 Å side length) will be different since the system does not contain one molecule any longer as shown in Figure 3.1c. In fact, the number density for a system of this size will be fluctuating as the molecule vibrates and it becomes impossible to determine a number density for such a system. Hence, a 1 m³ contained containing 1 mole of $CO_2$ gas is classical system (bulk system) and its behavior can be predicted by classical thermodynamic equations of state. In other words, the behavior of the system will be predictable by the general gas law. If we assume that carbon dioxide will behave as an ideal gas, then

$$pv = nRT$$

Knowing that $v$ for the system is 1 m³ and $n$ (the number of moles) is also equal to 1. Then, for our system

$$p \times 1(\text{m}^3) = 1 \text{ (mole)} \times 8.314472 \text{ (m}^3 \cdot \text{Pa} \cdot \text{mole}^{-1} \cdot \text{k}^{-1}) \times T \text{ (k)}$$

$$p = 8.314472 \times T \text{ (Pa)} \tag{3.7}$$

Eq. 3.7 enables the precise prediction of the system's pressure as a function of its temperature. It is crucial to note that such equation is *not applicable* for the other system we considered in this example because its number density $\rho(r)$ cannot be determined. In other words, the gas pressure inside a cubic container of the size of 3.52 Å *cannot be determined* using Eq. 3.7. In fact, even if the size of the contained was increased several times (edge size in the range of few nanometers), we still face the same problem and the gas pressure inside such a system is, still, undeterminable.

In Example 3.1, we saw how as the system gets smaller and smaller, our ability to define homogeneous thermodynamic functions, hence, formulate an equation of state for the system diminishes. Let's discuss another example where we follow what would happen as the system under consideration gets larger and larger.

**Example 3.2:** In this example, let's start with a system that is simply a car tire filled with air. The air pressure inside the tire is homogeneous and can be correlated to its temperature, still, using a modified form of the general gas law of the form:

$$P = znRT \tag{3.8}$$

where $z$ is a correction factor taking into consideration the deviation of air behavior from that of an ideal gas and can be found in many thermodynamic data tables. Now, let's allow the tire to expand and transforms into a spherical air shell with a thickness of 10 km resembling planet earth's atmosphere! The question, is atmospheric pressure within this system homogeneous? We all know that it is not. So, can we still use Eq. 3.8 to calculate atmospheric pressure? The answer is clearly no. Hence, we can conclude that a system with physical dimensions of a car tire is classical (bulk) while a system with physical dimensions of earth's atmosphere is not classical (nanoscale). While we can easily formulate equation of states capable of predicting the behavior of the first, we cannot do the same for the other so easily and other aspects must be considered.

In the previous two examples, using the number density thermodynamic function we showed that a thermodynamic system can transform from a bulk system into a nanoscale system as it becomes smaller or as it becomes larger. In both cases, we showed that the system transforms as such thermodynamic function (number density) becomes inhomogeneous. In the following example, we will investigate the performance of thermodynamic systems as other thermodynamic functions are concerned.

**Example 3.3:** Let us consider the case of the local internal energy density function $\phi(r)$ defined in Eq. 3.5. We know that the internal energy of a small sample of matter $(\delta U)$, indeed, depends on internal interactions among chemical species within the system volume $(\delta V)$. The system internal energy also depends on interactions with groups of such species which are outside the system boundaries.

Let's consider two cases to elucidate the effect of system size $(\delta V)$ on its thermodynamic treatment. We will, first, consider the simple case in which the strength of interactions within the system is on the same order as that with species outside the system. In this case, it is clear that the physical size $(\delta V)$ of the system plays a crucial role in determining the relative contribution of interactions with species outside the system volume to the system energy $(\delta U)$. It could be straightforward to realize that as the system volume increases, the out-side interactions relative contribution to the system energy are less relevant. Therefore, their effect on the ability to accurately determining the system energy $(\delta U)$ will diminish as the system physical size increases. Now, let us consider the more realistic case in which the strength of out-side interactions is substantially stronger than that of the within the system interactions. This could be due to an external field (mechanical, magnetic, electric, etc.) affecting the system of consideration or due to inhomogeneity in the system resulting from an interface or a membrane for example. This case represents most of the nanoscale systems that we could encounter, and hence, is essential to the field of nanotechnology. In such a case, the physical dimensions of the system $(\delta V)$ would play a much less significant role in determining the relative contributions to the system's energy $(\delta U)$. The system physical dimensions can be significantly large, and still its local thermodynamic functions are not homogeneous and cannot be uniquely determined, hence, its behavior cannot be accurately predicted. In other words, the internal energy of any thermodynamic system depends on two factors; the strength of interaction within the system, and the strength of the interaction between the system and its surrounding. If interactions within the system are stronger than its interaction with its surrounding, there will be a physical system size at which the surrounding effect is negligible and the system can be treated as classic or bulk. However, if the strength of the surrounding interaction is stronger than the strength of interactions within the system, such surrounding effects on the system will not diminish the internal energy of the system as its size increases; and regardless of the system's physical size, it cannot be treated as a classic or bulk system. It is important to note that in this example, we used the local internal energy $\phi(r)$ function as an example. The same argument applies to other local thermodynamic functions and intensity functions mentioned before.

To this end, it is important to note that the difference between bulk or classical thermodynamic systems that we are accustomed for and small, nanoscaled thermodynamic system that we are studying here is not the physical size of the system but the homogeneity of its local thermodynamic functions that are mainly pressure, temperature, chemical potential, and energy intensities.

## 3.2 Thermodynamics of Small Systems

As we have discussed before, our only theoretical tool to understand the physics of matter; thermodynamics and statistical mechanics, are derived for large, or macroscopic, systems, in which the fundamental thermodynamic functions of the system are homogeneous and determinable. Such tools become useless once the homogeneity condition is not met. This point is not new and was very well recognized and tackled by Terrell L. Hill, in 1962. Hill is considered the father of the thermodynamic of small systems. He was the first to point out the difference in thermodynamic treatment between large systems, where the number of molecules in the system ($N\rightarrow\infty$), and small systems, where the number of molecules is limited. The material system that dragged his attention in the early 1960s was colloidal particles and later in the late 1960s he applied his treatment to biological systems. It is crucial to note that Hill used the number of molecules ($N$) in his treatment as the sole indication of the system classification (large or small). While the number of molecules is directly related to the physical size of the system, we know by now that this is not a precise description for the system classification but it is important to know the stages of the development of nanoscaled systems.

Hill's aim of the proposed thermodynamic treatment was to extend the range of validity of classical thermodynamic definitions and interrelations into "nonmacroscopic systems" as he referred to them. It is important to note that the "nonmacroscopic systems" term used in the early 1960s is what we refer to as "nanosystems" nowadays. Another interesting point to realize is that Feynman's invitation to explore small systems announced in his famous lecture in December 1959 was already recognized, most probably independently, by Hill who submitted his first thermodynamic treatment of small systems in January 1962.

In his treatment of thermodynamic small system, Hill proposed to utilize "correction terms" in order to account for interactions which cannot be ignored as the system size hits the "small" boundaries. By this, Hill showed that thermodynamic equations can be generalized so that they will be valid for small systems.

To explain, let us consider a large one component system. Gibbs free energy ($G$) of such a system can be expressed as

$$G = Nf(p, T) \tag{3.9}$$

where $N$ is the number of particles (molecules) in the system, and $f$ is a function of pressure ($p$) and temperature ($T$) only. The chemical potential ($\mu$) of the molecules in such a system can be expressed as;

$$\mu = \left( \frac{\partial G}{\partial N} \right)_{p,T} \tag{3.10}$$

In the large thermodynamic system limit where $N \rightarrow \infty$, the system free energy $G \rightarrow Nf(p,T)$, and hence the chemical potential $\mu \rightarrow f(p,T)$, or in other words, becomes a function of pressure and temperature only. Now, if the system is small enough, Hill proposed the addition of correction terms to the original thermodynamic expression to ensure its validity. For example, the expression for Gibbs free energy should include such correction terms as

$$G = Nf(p, T) + a(p, T)N^{2/3} + b(T)lnN + c(p, T) \tag{3.11}$$

where $a$, $b$, and $c$ are functions of pressure and temperature.

Here, Hill added the correction terms $N^{2/3}$ to account for surface effects which are known to contribute to the free energy of the system (this is basically, the effect of system surroundings on the system as we discussed in Example 3.3), and $lnN$ to account for phonon confinement effects that is related to the case in Example 3.1, and a size independent term ($c$) to account for point contributions that is related to Example 3.1. He also pointed out that if line (or second order surface) contributions to the free energy of the system were to be accounted for, a new term of order $N^{1/3}$ should also be included. Also, if surface contributions to the free energy of a two-dimensional system (a single graphene sheet is a system of current interest for example) were to be considered, a correction term of the order $N^{1/2}$ should also be included. Hence, Eq. 3.11 can be generally written with as many correction terms as needed to fully account for all possible contributions. For example, an equation such as

$$G = Nf(p, T) + a(p, T)N^{2/3} + b(p, T)N^{1/3} + c(T)lnN + d(p, T) \quad (3.12)$$

is meant to account for volume, surface, line, phonon confinement, and point contribution to the system.

Assuming that the $f$, $a$, $b$, $c$, and $d$ functions are completely defined, and that they are strictly functions of pressure and temperature only, Eq. 3.12 can be used to describe thermodynamic functions in physically small systems as defined by a limited number of molecule ($N$).

Another major point of Hill's treatment of thermodynamics of small systems is pointing out that while the effect of the system's environment can be safely neglected in thermodynamic treatment of a large system, the same cannot be done for a small system. The environment of a thermodynamic small system plays a major role in its physical behavior and must be accounted for in any thermodynamic treatment of the system as we discussed in Example 3.3.

## 3.3 Thermal Fluctuations in Thermodynamic Systems

Thermal fluctuations and thermal stability are crucial points that need to be considered in the treatment of thermodynamic systems. This is the case discussed in Example 3.1. In a system containing $N$

particles, thermal fluctuations are usually on the order of $N^{(-1/2)}$, and hence, thermal stability is 1- fluctuations. For a large system where $N \rightarrow \infty$, thermal fluctuations are almost zero, and hence, the system is considered 100% thermally stable. However, for a small system of $N$ in the range of 100 atoms or molecules, a hundred molecules cluster of water for example; fluctuations are on the order of 10%, meaning that the system is only 90% thermally stable. For a water cluster or a metal particle of 25 molecules or atoms, the thermal fluctuations become on the level of 20% and, then, the thermodynamic behavior of such small system can be dominated by thermal fluctuations. Such phenomenon is, in fact, crucial for the future of electronic industry especially computer technology.

**Example 3.4:** Knowing that the state-of-the-art computer processors (i7) are 14 nm wide and that the size of a silicon atom is in the range of 0.2 nm. Calculate the thermal fluctuation effect for such processors.

**Solution:** Given that the width of the processor is 14 nm, and that the size of silicon atom is 0.2 nm, the processor is basically 70 atoms wide, and $N = 70$.

Thermal fluctuations are on the level of $\dfrac{1}{\sqrt{70}}$ which is 12%.

**Example 3.5:** What is the smallest processor size possible if the thermal fluctuations effect not to exceed 20%?

**Solution:** Thermal fluctuations = $\dfrac{1}{\sqrt{N}}$ = 0.2. Hence, $N = 25$ atoms.

Hence the smallest processor size = $25 \times 0.2$ (nm) = 5 nm.

## 3.4 Configurational Entropy of Small Systems

For large systems that would mix ideally at certain temperature and pressure, the driving force for mixing is the configurational entropy. This is a well-known natural phenomenon that takes place in solids (copper and gold), liquids (water and methanol), and gas (oxygen and nitrogen) bulk systems. In such cases, the two components in the system, as we all know and expect, would mix to maximize the system's configuration entropy, hence reaching the state of equilibrium as we observe it. The interesting question would be "would these components mix in a small system configuration as they do in a large system configuration?" It is widely known that based on statistical mechanics concepts, the state of equilibrium for a system is the most probable state of the system. Configurational states of a system are just the different configurations the system can assume. In other words, these are the different ways the system can arrange its components while maintaining equilibrium (as defined by the minimum free energy state) under the effect of constant, temperature, pressure, and other thermodynamic fields. In order to address our question, let us consider the simplest case possible for mixing as controlled by configurational entropy. A system of four atoms of component ($a$) and four atoms of component ($b$) with an interface as shown in Figure 3.2. The number of ways ($\Omega$) such eight atoms can be distributed among the available 8 spatial positions can be calculated according to

$$\Omega = \frac{N!}{n_a!n_b!} \tag{3.13}$$

where $n_a$ and $n_b$ are the number of species a and b, respectively, and $N = n_a + n_b$ is the total number of atoms or molecules in the system which is 8 in our case.

The first configuration of the system to consider is the configuration where the atoms are not mixing and each species is isolated on one side of the interface. In this case the four "a" atoms are on one side and the four "b" atoms are on the other side of the interface. The number of ways the system can assume this configuration is clearly 1. We will express this as $\Omega_{4:0} = 1$. If one of the "a" species crosses the interface, note that this will require that one of the "b" species also crosses the interface in the opposite direction, it can be shown that this partial mixing situation configuration of the system can be assumed by 16 different ways, and hence, $\Omega_{3:1} = 16$. Now let us consider the case when two of the "a" species cross the interface to generate the total mixing situation where the composition will be homogeneous across the system. In this case, it can be shown that the number of spatial distributions of the atoms among available spatial configurations can be 36, so $\Omega_{2:2} = 36$. Similar calculations can be reached for the remaining two configurations, $\Omega_{1:3}$ and $\Omega_{0:4}$.

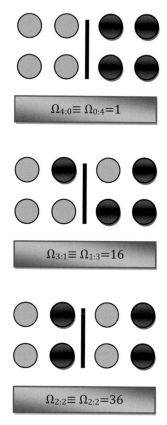

**Figure 3.2** Representation of two small systems in contact.

Now let us address the equilibrium state issue, which of the aforementioned five configurations would be the most probable configuration that the system would assume? Well, the total number of ways the system can distribute its 8 constituents among the available 8 spatial positions turned out to be 1 + 16 + 36 + 16 + 1 = 70. This means that the probability of the five different configurations we discussed should be 1/70, 16/70, 36/70, 16/70, and 1/70, respectively. This indicates that the two unmixed configurations would have a probability of 1.4% each, the two partially mixed configurations would have a probability of 22.8% each, and the totally mixed configuration would have a probability of 51%. The totally mixed situation (homogeneous composition) in this case is the *most probable state* of the system. However, it is not the *only possible state* of the system. It is clear from our example that a system made of only 8 atoms would spend only 51% of its time in the totally mixed state and the other 49% of its time will be in the partially or unmixed states.

If the system is large in size $N \rightarrow \infty$, the *most probable state* becomes *the only possible state*. Hence, the system would spend 100% of its time in such a state which we refer to as the state of equilibrium.

This is why if we lay a typical object ($N$ is on the order of $10^{23}$) on the table, the object stays there as long as no other external field causes it to move. It remains in its only possible state. *For a small system, however, with limited number of atoms, equilibrium is truly the most probable but not the only possible state.* Other non-equilibrium states are possible too and, hence, it is possible, according to statistical mechanics principals that a theoretical molecular assembly made of limited number of molecules, as the one shown in Figure 3.3, would suddenly reverse its rotation direction or separate and assume a different configuration.

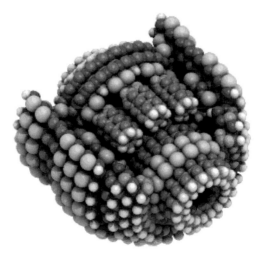

**Figure 3.3** This is the MarkIII(k), a nanoscale planetary gear designed by K. Eric Drexler. A planetary gear couples an input shaft via a sun gear to an output shaft through a set of planet gears (attached to the output shaft by a planet carrier). The planet gears roll between the sun gear and a ring gear on the inner surface of a casing. This and other molecular assemblies are available on the web at: http://nanoengineer1.com/content/index.php?option=com_content&task=view&id=40&Itemid=50

> In nanoscaled systems, equilibrium is the most probable state. In bulk system, the most probable state is the only possible state.

## 3.5 Molecular Machinery

The discussions in the previous sections by no means should mislead the reader into believing that molecular assemblies are not thermodynamically possible or stable. Nature has been designing and utilizing very successful and efficient *molecular machinery* for millennia. Bacteria and green plants run the *nanomachinery* that makes life possible using the very same clean, sustainable, and readily available source of power that we are still trying to harness—sunlight. Other forms of natural molecular machines utilize other forms of energy to function. Humans are good example for fuel diversity. A number of studies and reviews have described the concept and first principals of designing molecular machines or nanomachines similar to those perfected by nature. Our previous discussion was meant to indicate that designing and utilizing nanostructured systems referred to as *molecular machinery* necessitates a deep consideration, appreciation, and understanding of the thermodynamic of such small systems. The subject of nanosystems is too sophisticated to be considered from a single discipline viewpoint. The new frontier of nanotechnology is interdisciplinary in nature and any advances in such a field requires collaboration and deep appreciation of physics, chemistry, materials science, biology, in addition to a number of different disciplines of engineering. It is crucial to realize that nature in its 3.5 billion years quest to design efficient molecular machinery; it designed molecular machines capable of harnessing Gibbs free energy changes and transforming such changes into any of the other forms of energy including mechanical work. Different nanomachines designed by nature can harness changes in the Gibbs free energy of the system resulting from changes in any of the thermodynamic potentials affecting the system. These include changes in mechanical force or pressures, changes in length or volume, changes in temperature, changes in chemical potential or numbers of a chemical species, changes in the oxidative state or numbers of electrons of chemical moieties, changes due to the absorption of electromagnetic radiation, or even changes in the extent of spatial order (the entropy) of the system. Figure 3.4 demonstrates all possible (observed and supposed) energy conversions that living organisms (molecular machines based upon protein molecules) can perform. The bold arrows indicate the energy conversions that have been observed.

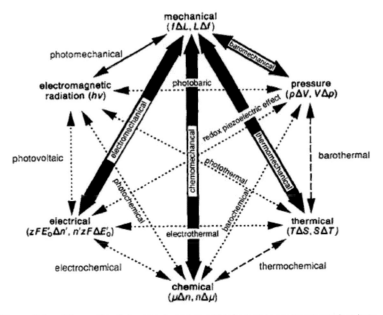

**Figure 3.4** All possible (observed and possible) energy conversions that living organisms (molecular machines based upon protein molecules) can perform. The bold arrows indicate the energy conversions that have been observed. Reproduced with kind permission from D. Urry, *Angew. Chemie*, International Edition in English, 1993, 32, 819. Copyright VCH, 1993.

It is interesting to realize that natural polymer molecules (in the form of proteins) are not the only moieties capable of assembling molecular machines. Certain manmade polymers and other smaller molecules can be utilized as well. With so many energy conversion possibilities, tremendous building blocks, and a deep understanding of the physics of small systems, nanotechnology can definitely benefit humankind if wisely utilized.

To conclude, it can be seen that in nanostructured systems, forces that are usually have negligible effects on the physics of the system become dominant and, basically, control the system's behavior leading to the observation of new, and unusual, physical phenomena. In addition, we briefly discussed the principals and basic concepts of nanomachines. The important message to remember here is that nanomachines, as a product of nanotechnology, are not about making miniature versions of our machines based on the same concepts of design we currently utilize. In fact, nanomachines should be based upon concepts utilized by nature, the same thermodynamic concepts determining the behavior of small systems. In the following section, we will give a number of examples to demonstrate different physical phenomena observed for typical material systems once they approach the nanodomain.

## Problems

1. What are our current theoretical tools to understand the behavior of matter?
2. Define the meaning of "equation of state"?
3. At thermodynamic equilibrium of a bulk system, thermodynamic functions must be homogeneous. Explain this statement with examples.
4. What is the basic assumption of classical thermodynamic?
5. Explain the quasi-thermodynamic assumption.
6. Define the following thermodynamic functions:
   a. Local number density
   b. Local chemical potential
   c. Local pressure
   d. Local internal energy
7. For the system to be accurately treated as bulk system, the local thermodynamic functions of the system must be ---------. (complete the statement).
8. We must treat a thermodynamic system as a small system once its local thermodynamic functions are no longer homogeneous within the system. Explain this statement.
9. A thermodynamic system can turn into a small system only as its physical size is decreased. State if this statement is true or not, and explain your answer.
10. Explain Hill's approach for the thermodynamic treatment of small systems.

11. System surrounding conditions can be ignored in both large and small thermodynamic systems. State if this statement is correct or not, and explain your answer.
12. Explain the concept of thermal fluctuations.
13. Define thermal and configurational entropy.
14. What is the difference in equilibrium in nanoscaled and bulk systems?
15. State three energy conversion mechanism that living organism can perform.

# Chapter 4

# Scales of Thermodynamic Inhomogeneity

## 4.1 Introduction

So far, it is clear that thermodynamic functions and equations of state are our most powerful and only tools to understand and be able to predict the behavior of matter. Such tools, however, can only be applied in cases where the thermodynamic functions can be homogeneous. As we have seen in the previous chapter, thermodynamic function homogeneity condition can be lost as the system physical size becomes smaller or larger. Hence, a need to set conditions for the system behavior (classic or nanostructured) and relate its properties to its physical size aroused. In other words, the questions to ask would be: Is there a condition, or a set of conditions, that once met the system becomes non-classic and must be treated as a nanostructured system? The answer for such questions is "yes." And that is the topic of this chapter.

## 4.2 Scales of Thermodynamic Inhomogeneity

Several, namely five, characteristic scale lengths have been identified and shown to affect the nature of thermodynamic behavior of a system. If the physical size of the thermodynamic system in

*Gigantic Challenges, Nano Solutions: The Science and Engineering of Nanoscale Systems*
Maher S. Amer
Copyright © 2022 Jenny Stanford Publishing Pte. Ltd.
ISBN 978-981-4877-74-9 (Hardcover), 978-1-003-14704-6 (eBook)
www.jennystanford.com

consideration is on the order of any of these scale-lengths, the system cannot be treated as classical thermodynamic system and must be treated as a nanostructured system if accurate prediction of its behavior is to be achieved. These scale-lengths are thermal gravitational length scale, capillarity length scale, Tolman length scale, line tension ($\tau$), and the line tension to surface tension ($\tau/\sigma$) ratio length scale, and the correlation length scale.

## 4.2.1 Thermal Gravitational Length Scale

Since all systems of interest on planet earth are under the effect of its gravitational field and are also affected by its temperature (thermal field), it is important to realize the restriction such gravitational and thermal fields might impose on the use of classical thermodynamic. For a *one phase* system, the characteristic length associated with the combination of a thermal and gravitational fields ($l_g$) is given by

$$l_g = \frac{kT}{mg} \tag{4.1}$$

where $T$ is the system temperature in Kelvin), $k$, is Boltzmann constant, $m$ is the molecular mass, and $g$ is the gravitational field constant.

This means that if the physical size of the system under consideration is much smaller than its thermal gravitational length, the system can be treated classically with an average temperature and pressure. However, if the system is on the length-scale of its thermal gravitational length, it must be treated as a nanostructured system. This means that its local thermodynamic functions must be defined locally and will be functions of local temperature and density:

$$p(r) = p[T(r), \rho(r)], \text{ and } \mu(r) = \mu[T(r), \rho(r)] \tag{4.2}$$

**Example 4.1:** Calculate the thermal gravitational length on planet earth of a gas mixture of nitrogen and oxygen of 78% and 22% composition, respectively, at ambient temperature and pressure conditions.

**Solution:** Here we have a straight forward problem of a gas mixture 78% $N_2$ and 22% $O_2$ at ambient pressure (1 atm) and temperature (298 K), and we need to calculate the system's thermal gravitational length ($l_g$) on planet earth ($g = 9.81$ m/s$^2$). The equation to apply in this case is

$$l_g = \frac{kT}{mg}$$

The molecular weight for $N_2$ is 28 g/mole and that for $O_2$ is 32 g/mole.

Hence, the molecular weight for our mixture is

$$m = 0.78 \times 28 + 0.22 \times 32 = 28.88 \text{ g/mole}$$

$$l_g = \frac{(1.38 * 10^{-23} \text{ m}^2\text{kg s}^{-2}\text{K}^{-1})(298 \text{ K})}{\left(28.88 \frac{\text{g}}{\text{mole}}\right)\left(9.81 \frac{\text{m}}{\text{s}^2}\right)\left(0.001 \frac{\text{kg}}{\text{g}}\right)\left(\frac{1}{6.023 \times 10^{23}} \text{mole}\right)}$$

$$l_g = 8742.6 \text{ m}$$

Compare this value to the 10,000 m, which is the thickness of the troposphere! (hint: the troposphere is the lowest layer of planet earth's atmosphere)

Example 4.1 shows us the planet earth's troposphere (the air layer around earth) has a thickness on the same size of air's thermal gravitational length scale at ambient temperature. This immediately tells us that to predict the behavior of the troposphere (as a system) the local thermodynamic functions (i.e., $T_{(r)}$, $P_{(r)}$, $\mu_{(r)}$, etc.) must be determined and used to formulate any equation of state capable of predicting this system's behavior (this is known as weather forecast).

## 4.2.2   The Capillary Length

If we consider a *two-phase system*, a liquid and a gas, in equilibrium under a gravitational field, the liquid phase will be at the bottom and the vapor will be above it with an interface in between. Due to thermal fluctuations, the interface will instantaneously depart from planarity (being planar), creating what is referred to as *capillary waves.* The characteristics of these capillary waves depend substantially on the system size and the strength of the gravitational field and the surface tension of the two phases. The characteristic length that governs these capillary waves is known as *the capillary length* ($l_c$) and is defined as

$$l_c^2 = \frac{2\sigma}{g(\rho_l - \rho_g)} \tag{4.3}$$

where $\rho_l$ and $\rho_g$ are the (mass) density of the liquid and gas phases, respectively (kg/m$^3$), and $\sigma$ is the liquid surface tension (J/m$^2$).

Capillary wavelength is typically in the range of 10$^{-3}$ m. For water/air interface at 0 °C in earth's gravitational field, it is 3.93 mm. Hence, water surface waves with wavelength much larger than the capillary length (for example, sea waves) will be completely governed by gravity and the wind mechanical forces. Water surface tension will have no effect on the water surface in this case. As the surface physical dimensions, however, become smaller and smaller till they reach a size that is comparable to the capillary length (on a scale of millimeters), the surface will no longer be planar but will take the shape of a wave or a curve as shown in Figure 4.1.

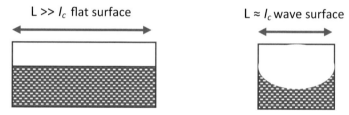

**Figure 4.1** Liquid/gas surface at different system length scales.

Now that the liquid surface is curved, the surface tension forces ($2\pi r\sigma$) will have a vertical component ($2\pi r\sigma\cos\theta$) pulling the liquid up as shown in Figure 4.2. This will cause the liquid to rise to the height ($h$) above the liquid surface in the surrounding. Such phenomenon is known to us as *capillarity*. Capillarity was first investigated by Leonardo da Vinci in the early fifteenth century and was also the subject of Einstein's first real paper in 1901. At equilibrium, the surface tension forces ($2\pi r\sigma\cos\theta$) must equal the weight of the liquid column ($W = \pi r^2\rho gh$) and, hence, the height ($h$) of the liquid column inside a tube of radius ($r$) can be expressed as

$$h = \frac{2\sigma\cos\theta}{\rho g r} \tag{4.4}$$

For water in a glass tube at ambient conditions, Eq. 4.4 can be reduced to

$$h \approx \frac{14 \times 10^{-6}}{r} \text{ m} \tag{4.5}$$

where $r$ is in meters. Hence, if the tube internal diameter is in the meter range, $h$ will be in the micrometer range, and the phenomenon is hardly observed. Once the tube internal diameter is reduced into the millimeter range, $h$ will be in the centimeter range, and the phenomenon becomes very observable. It is important to note that the capillary phenomenon is essential for life and it is the reason water rises up tall trees carrying nutrition to maintain tree life and growth. Feeding tubes on the order of few microns are needed to guarantee feeding of 10 m tall trees.

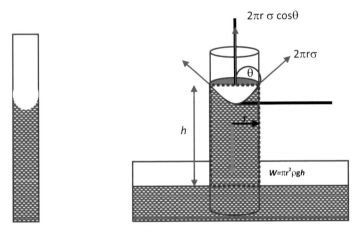

**Figure 4.2**    Capillarity effect at liquid/gas interface.

### 4.2.3    Tolman Length Scale

A point that should be of extreme interest in the field of nanofluids is the applicability of Laplace's equation in situations where the radius of a liquid droplet in gas medium or gas bubble in a liquid medium is in the nanoscale. According to Laplace's equation, the excess pressure inside a liquid droplet or a bubble ($\Delta p = p^1 - p^g$) can be expressed as

$$\Delta p = p^l - p^g = \frac{2\sigma}{R} \tag{4.6}$$

where $R$ is the radius of the gas bubble or liquid droplet and $\sigma$ is the surface tension between the gas and liquid. Laplace treated the surface tension as a material constant, at constant temperature, that is not a function of geometry, which means

$$\sigma = f(T) \neq f(R) \qquad (4.7)$$

Tolman, however, showed that the surface tension ($\sigma$) decreases with decreasing the droplet radius over a wide range of circumstances and that it can be expressed as

$$\sigma_R = f(R) = \sigma_\infty \left( 1 - \frac{2\delta}{R} + \ldots \right) \qquad (4.8)$$

where $\sigma_R$ is the surface tension of a droplet or a bubble of a radius $R$, and $\sigma_\infty$ is the surface tension of a planar surface. Here, $\delta$ is the *Tolman length*. It is important to emphasize that in Eq. 4.8, the Tolman length is defined as a coefficient in an expansion of $1/R$ and, therefore, is not a function of $R$.

Tolman length has received considerable theoretical attention. However, there are still some issues that remain completely unresolved. While there is some agreement in the literature that it is not a sharp dependent of temperature, a huge controversy regarding its value does exist. It has been shown recently that the value of Tolman length sensitively depends on the interaction potentials in the liquid. The discrepancy in its sign and its dependence on the interaction potential is still not well understood. Most widely accepted value for Tolman length ranges between 1 nm and a fraction of a nanometer. While the exact value of Tolman length has been a controversy for the past decades, recent computer simulations based on molecular dynamics and density function theory showed that the Tolman length ($\delta$) can be estimated from the liquid surface tension ($\sigma$) (N/m) and the bulk liquid isothermal compressibility ($\beta$) (Pa$^{-1}$) according to the equation;

$$\delta \approx -\beta\sigma \qquad (4.9)$$

It is important to note that a negative Tolman length means that the surface is curved toward the liquid and a positive Tolman length means that the surface is curved toward the gas.

Tolman equation can yield an expression for the excess pressure inside a liquid droplet ($\Delta p$) as

$$\Delta p = \frac{2\sigma}{R}\left(1 - \frac{2\delta}{R} + ...\right) \tag{4.10}$$

with the first term being the same as Laplace's equation.

Or, the pressure inside the liquid droplet can be calculated using a modified Laplace equation as

$$\Delta p = p^l - p^g = \frac{2\sigma_R}{R} \tag{4.11}$$

The important point to realize here is that the surface tension, as a physical property of a liquid, cannot be uniquely defined as the droplet radius is on the order of Tolman length. This is a very clear example of the effect of entering the nanodomain on our understanding of materials behavior. A simple physical property, such as surface tension, first investigates by Leonardo da Vinci in the early fifteenth century and officially introduced by J. A. Segner in 1751, becomes completely unexplored and ununderstood once the system size becomes of the order of Tolman length. That is 1 nm or less.

**Example 4.2:** Given that the surface tension of water at room temperature is 70 mN/m, and that its isothermal compressibility at that temperature is 0.46 GPa$^{-1}$, determine the pressure inside a 100 nm diameter water droplet in air.

[Hint: consider atmospheric pressure equal to zero]

**Solution:** According to equations 4.11, and since we will consider the atmospheric pressure as our datum with a value of zero, the pressure inside the water droplet can be calculated as

$$\Delta p = p^l - p^g = p^l - 0 = \frac{2\sigma_R}{R}$$

The Tolman length can be determined from Eq. 4.9 as

$$\delta \approx -\beta\sigma = 0.46 \times 10^{-9}\left(\frac{m^2}{N}\right) \times 70 \times 10^{-3}\left(\frac{N}{m}\right)$$

$$\therefore \qquad \delta = -32.2 \times 10^{-12}\,(m) \quad or \quad -0.322\,(\text{Å})$$

The surface tension of a water droplet with diameter = 100 nm ($R$ = 50 nm) can then be calculated as

$$\sigma_R = \sigma_\infty \left( 1 - \frac{2\delta}{R} \right)$$

$$\sigma_{50nm} = 70 \times 10^{-3} \left( \frac{N}{m} \right) \left( 1 + \frac{2 \times 32.2 \times 10^{-12}(m)}{50 \times 10^{-9}(m)} \right)$$

$$\sigma_{50nm} = 70.09 \times 10^{-3} \left( \frac{N}{m} \right)$$

$$\therefore \quad p^l = \frac{2\sigma_R}{R} = \frac{2 \times 70.09 \times 10^{-3} \left( \dfrac{N}{m} \right)}{50 \times 10^{-9}(m)} = 2.8 \times 10^6 \left( \frac{N}{m^2} \right)$$

$$\therefore \qquad p^l = 2.8 \text{ MPa}$$

It is important to note here that because the water compressibility is so low, the Tolman length was very small. In addition, since the droplet size in this example is very large compared to the Tolman length (almost 3 orders of magnitude), the surface tension of water was not much affected.

The situation is different with more compressible oils as we will see in the next example.

**Example 4.3:** Calculate the pressure required to pass oil with compressibility = 1 Pa$^{-1}$ and surface tension = 34 mN/m through an orifice that has a diameter of 100 nm.

**Solution:** It should be noted that to pass a liquid through an orifice a pressure must be applied to the liquid. The applied pressure should be equal to the pressure inside a droplet of the same size.

$$\therefore \qquad \delta = -\beta\sigma = -34 * 10^{-3} \text{ (m)} \quad or \quad -34 \text{ (mm)}$$

In such a case, a 100 nm diameter droplet will have a surface tension of;

$$\sigma_{50nm} = 34 \times 10^{-3} \left( \frac{N}{m} \right) \left( 1 + \frac{2 \times 34 \times 10^{-3}(m)}{50 \times 10^{-9}(m)} \right)$$

$$\therefore \qquad \sigma_{50nm} = 2.3 \times 10^6 \left( \frac{N}{m} \right)$$

This means that the pressure inside this oil droplet or the pressure needed to pass this will be through a 100 nm diameter orifice is

$$\therefore \quad p^l = \frac{2\sigma_R}{R} = \frac{2 \times 2.3 \times 10^6 \left( \dfrac{N}{m} \right)}{50 \times 10^{-9} (m)} = 92 \times 10^{12} \left( \frac{N}{m^2} \right)$$

$$\therefore \qquad p^l = 92 \text{ TPa}$$

This is an extremely high level of pressure and cannot be practically achieved.

## 4.2.4 Line Tension ($\tau$) and the ($\tau/\sigma$) Ratio

Another important area of nanotechnology covers self-assembly and formation of molecular monolayer films known as Langmuir–Blodgett films. The technique depends on the spreading of a substance, usually a polymer or nanoparticles in a solvent, on the surface of water to form a continuous ultra-thin film. Depending on the nature of spread solution, it may spread and form the desired film or may not spread and form a set of liquid lenses on the surface of the water. Thermodynamic investigation of the spreading (wetting) and non-spreading (non-wetting) regimes revealed another important scale of inhomogeneity that is crucial for such a process. This length scale is the *line tension* ($\tau$) to the surface tension ($\sigma$) ratio. Let us further clarify this point.

If three phases, $\alpha$, $\beta$, and $\gamma$ (in our case these three phases would be the water substrate, the spread solution, and air), meet at three surfaces, and if the three surfaces meet at a line, then the free energy ($F$) of such a system can be expressed as

$$F = V^\alpha \psi^\alpha + V^\beta \psi^\beta + V^\gamma \psi^\gamma + A^{\alpha\beta} \sigma^{\alpha\beta} + A^{\alpha\gamma} \sigma^{\alpha\gamma} + A^{\gamma\beta} \sigma^{\gamma\beta} + L^{\alpha\beta\gamma} \tau^{\alpha\beta\gamma}$$

$$(4.12)$$

where $V$ denotes the phase volume measured to the equimolar dividing surface, $A$ denotes the contact area between the phases, $L$ is the length of the three phases contact line, and $\tau^{\alpha\beta\gamma}$ is the line tension. It is important to note that unlike the surface tension which is always positive, the line tension can be positive or negative.

The ratio ($\tau/\sigma$) is very important for the physics of small systems. For soap solutions, for example, this ratio is on the order of 20 nm. Hence it is clear that for droplets or bubbles with dimensions between 1 mm and 1 µm, the physics of the system is mainly determined by surface tension and can be handled by ignoring both gravity and line tension. However, if the system dimensions are on the order of the $(\tau/\sigma)$ ratio (20 nm or less) the effect of line tension cannot be disregarded and has to be accounted for in any attempt to understand the behavior of the system.

Another important length scale to consider, especially while dealing with nanoparticles, is the surface tension ($\sigma$, in joules/m$^2$), to the bulk energy density ($\phi$, in joules/m$^3$) ratio ($\sigma/\phi$). Such a ratio has a length scale usually in the nanometer range, or less, depending on the nature of the material. For example, the ratio for argon at its boiling point is in the range of 0.5 nm. The ratio for water at 25 °C, drops down to about 0.3 Å.

Talking about line contributions, we can still consider the case where multiple lines (usually three as in the triple point in foams) of phase contacts meeting in a point. This can hardly be considered tension. However, it contributes to the system's free energy independent of the system size.

## 4.2.5    The Correlation Length ($\xi$)

In any system, regardless of its physical size, fluctuations in one region of the system can influence other regions in the system. In such a situation, the two regions of the system are said to be *correlated*. The *correlation length* ($\xi$) is the distance or the range over which different regions, or points, in the system can be correlated. This means that, if two points in the system are separated by a distance equal to or smaller that the system's correlation length, then these two points will influence each other and any fluctuations or changes in one point will be felt by the other. However, if the two points are separated by a distance that is larger than the system's correlation distance, then the two points will not *feel* each other and each of them will behave as if the other does not exist. This takes us back to the effect of external perturbation fields (mechanical, thermal, magnetic, etc.) on the behavior of materials systems. It is important to note the different effects of external perturbation field on the core atoms in a particle that much larger than the correlation length and that in a particle with a physical size comparable to the correlation length. This is illustrated in Figure 4.3. If we assume that the external perturbation field is a thermal field, in the particle with size scales to the correlation length (Figure 4.3 a), the effect of the thermal field will be "felt" simultaneously on both sides of the particle since they are within the correlation length from each other. In Figure 4.3b, however, where the two sides of the particle are far from each other, they are not correlated and the heat from the external thermal field will be "felt" by the far side of the particle after a while depending on a materials constant of the particle material which is its thermal diffusivity ($\alpha$). Thermal diffusivity is a materials constant that is defined as

$$\alpha = \frac{k}{\rho c_p} \left( \frac{\mathrm{m}^2}{\mathrm{s}} \right)$$

where $k$ is the materials thermal conductivity (W/m·K), $\rho$ is the mass density (kg/m$^3$), and $c_p$ is the specific heat capacity under constant pressure (J/kg·K). This is very important because the phenomenon controls how fast we can heat or cool a materials system. If the system size scales to its correlation length, it means that all parts of the system are correlated and, hence, there will be no limit on how fast we can heat or cool the system (add or take heat from it) regardless of the system's thermal diffusivity. However, of the system size is much larger than its correlation length, then the system thermal diffusivity becomes the limit and the rate at which heat can be added or withdrawn from the system will be limited by the system's thermal diffusivity.

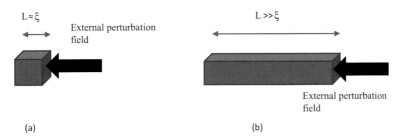

**Figure 4.3** External perturbation field acting on a particle with a size (a) that scales to the material correlation length, and (b) that much larger than the material correlation length.

In addition, it is important to note that while correlation length in materials (on the atomic scale) tends to be in the range of 1 to few nanometers, point correlations in systems do exist regardless of the size of system. For example, correlations among the planets in our, and other, solar systems do affect the planets trajectories with gravitational correlation length on the order of few million miles. Tidal phenomenon on our planet is a direct result of the correlation between earth and its moon. Correlations in the distribution of galaxies were also observed and found to provide some important clues regarding the structure and evolution of the Universe on scales larger than individual galaxies.

As we mentioned earlier, in a material system correlation length, or molecular interaction range, is typically considered to be on the order of 1 nm. Recent studies, however, showed that correlation length in certain systems can extend in the range of 7 ~ 10 nm. The correlation length ($\xi$) in materials was also found to depend on the temperature ($T$) and to diverge at the system's critical temperature ($T_c$). Mathematically, it can be expressed as

$$\xi \sim |t|^{-\nu} \text{ and } t = \frac{T - T_c}{T} \tag{4.13}$$

where $\nu$ is known as the *critical exponent*.

Divergence of the correlation length at the critical temperature means that the correlation length becomes very large. Therefore, very far apart molecules become correlated, and the system's physical behavior, regardless of the system size, becomes completely different from what we are accustomed to.

Away from the critical point, where the correlation length extends to reach the physical dimension of a system, if we reduce the physical dimensions of a system to match its correlation length, then the same phenomenon can be observed. For a system in such state, thermodynamic local functions cannot be uniquely defined. Hence, the system behavior cannot be mathematically expressed or predicted based on classical thermodynamic or statistical mechanics principals.

## 4.3   Summary

To this end, we have seen that our current understanding and ability to predict the physics of materials systems is based upon our ability to formulate thermodynamic equations of state for such systems. We have also discussed the fact that once thermodynamic inhomogeneity exists in a system at certain length scales, the system has to be treated as a thermodynamic small system that requires the definition of local thermodynamic functions. Most importantly, we have discussed different scales of inhomogeneity at which local thermodynamic functions cannot be uniquely defined. We have also shown that the physics of a system depends on where the system size lies in respect to these characteristic length scales. We find that

$$\xi \sim \sigma/\phi \sim \delta \sim \tau/\sigma \sim 1 \text{ nm} \tag{4.14}$$

Once the system physical size is on the order of its correlation length, the system is thermodynamically small, or in other words, the system is nanostructured.

It is also interesting to realize that the correlation length in the universe is on the order of millions of light years rendering the universe a nanostructured system! A nanosystem is not necessarily small after all.

## Problems

1. What are the five different homogeneity lengths?
2. Define the thermal gravitational length scale.
3. Once the physical size of the system scales to one of the homogeneity lengths, the system is considered -----------. (complete the sentence)
4. Define the capillary length scale.
5. State Laplace's equation.
6. Define Tolman's length scale.
7. While Laplace treated surface tension as ------------, Tolman showed that surface tension does depend on -----------. (complete the statement)
8. Estimate the Tolman length for:
   a. Water at 300 K
   b. Methanol at 300 K
   c. Mercury at 300 K
9. Define the correlation length scale.
10. What is the value of correlation length scale for matter on planet earth?

11. Correlation length scale is always very small. Is the statement correct? Explain your answer.

12. Correlation length scale is a materials constant which is independent of temperature. Is the statement correct? Explain your answer.

# Chapter 5

# Depletion Forces and Surface Tension Effects

## 5.1 Entropic and Depletion Forces

One of the most important thermodynamic functions for a nanoscale system is its entropy. The total entropy of a system ($S_{total}$) is basically the sum of two different types of entropy; thermal entropy ($S_{th.}$) and configurational entropy ($S_{conf.}$). The thermal entropy depends on the number of ways the energy can be distributed among its constituents ($\Omega_{th.}$), and configurational entropy depends on the number of ways the system constituents can be distributed within the available system space ($\Omega_{Conf.}$). This can be expressed mathematically as follows;

$$S_{total} = S_{th.} + S_{conf.} = k_B \ln\Omega_{th.} + k \ln\Omega_{conf.} \tag{5.1}$$

where $k_B$ is Boltzmann's constant.

Another way to calculate the configurational entropy is to assume ideal mixing (no enthalpic interactions) between the system constituents and hence,

$$S_{conf.} = -R'(X_A \ln X_A + X_B \ln X_B) \tag{5.2}$$

where $X_A$ is the volume fraction of constituent A, and $X_B$ is the volume fraction of constituent B. $R'$ is the universal gas constant, expressed in J/K units = $8.314 \times$ number of moles in the system.

*Gigantic Challenges, Nano Solutions: The Science and Engineering of Nanoscale Systems*
Maher S. Amer
Copyright © 2022 Jenny Stanford Publishing Pte. Ltd.
ISBN 978-981-4877-74-9 (Hardcover), 978-1-003-14704-6 (eBook)
www.jennystanford.com

For a nanoscale system at constant temperature, it is actually the configurational entropy that we are interested in. Without going into mathematical derivations, it is very simple to realize that as the volume available for the system's constituents increases, the number of ways they would distribute themselves also increases, and hence, the configurational entropy of the system increases. Now, let's consider Figure 5.1. The figure depicts a two-dimensional area with a two-dimensional disc of radius $r$ in it. Focusing on the center of the disc, the area available for the center is shown in Figure 5.1a. The shaded area is not available, or in other words excluded, from the system area since the disc center cannot be located within such area. If a second disc is added to the system. Figure 5.1b shows the excluded area. Note that around each disc, an additional excluded area is added since the center of one disc cannot be located in such area otherwise the two discs will overlap which is not physically possible. Let's not consider Figure 5.1c, where the two discs are touching each other. In such a case, the two excluded areas around the discs overlap, hence, the total excluded area in the system increases. This means that the total available area in Figure 5.1c is larger than the total available area in Figure 5.1b. Hence, the configurational entropy of the system arrangement shown in Figure 5.1c is higher than the configurational entropy of the system shown in Figure 5.1b. Such increase in the system's entropy as its constituents agglomerate is the driving force for agglomeration known as the entropic or depletion force.

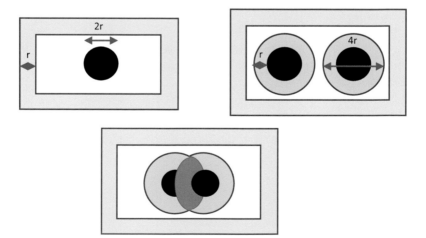

**Figure 5.1** (a) Excluded area for a single particle in a box, (b) excluded area for two particles in a box, and (c) excluded area for two particles in a box when the particles agglomerate.

**Example 5.1:** Arrange two circular discs in a box with maximum configurational entropy.

**Solution:** Since the maximum configurational entropy arrangement is required, then the excluded area must be maximized. This requires maximum possible overlap among all excluded areas. Figure 5.2 shows such arrangement.

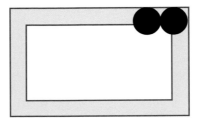

**Figure 5.2** Arrangement with maximum possible overlap among all excluded areas.

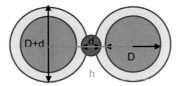

**Figure 5.3** Arrangement considered in Asakura–Oosawa model. $D$ is the diameter of large spheres, $d$ is the diameter of small spheres, and $h$ is the center to center distance between the large spheres.

The quantification of the depletion force has been in the focus of many theoretical investigations since the 1950s. The first model developed to quantify the depletion forces was the Asakura–Oosawa model. In their model, Asakura and Oosawa calculated the depletion force in large rigid spheres dispersed in a liquid made of small spheres. Figure 5.3 shows the arrangement considered in the model. The depletion force (F) can then be calculated as;

$$F = 0 \text{ for } h \geq d + D \tag{5.3}$$

$$F = -\frac{N}{4V}k_{B}T\pi(D+d+h)(D+d-h) \quad \text{for} \quad h < D+d \tag{5.4}$$

where $N$ is the total number of small spheres, $V$ is the system volume, $D$ is the diameter of large spheres, $d$ is the diameter of small spheres, $k_{B}$ is Boltzmann's constant, $h$ is the center to center distance between the large spheres, and $T$ is the absolute temperature of the system. The negative sign indicates that the force is an attractive force between the two large spheres.

It is important to note that the depletion force does depend on the particle shape and size. Figure 5.4 depicts the depletion potentials calculated for different shape constituents in the mixture. It is clear from the figure that the weakest potential is between two spheres, and the strongest potential is the one between two plates.

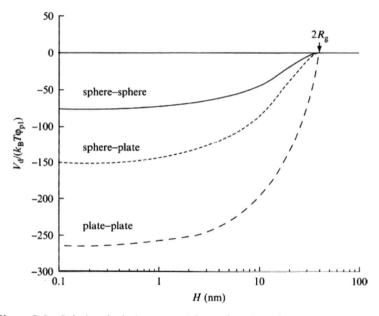

**Figure 5.4** Relative depletion potential as a function of separation distance calculated using Asakura–Oosawa model for two large spheres, two plates and a large sphere and a plate immersed in a fluid of small hard spheres. Small sphere radius $R_g = 20$ nm. The radius of large sphere is 103 nm and the surface of plates is 150 × 150 nm. $H$ is the distance between the surfaces of large bodies. Reproduced with kind permission from J. W. Goodwin, *Colloids and Interfaces with Surfactants and Polymers: An Introduction*, Copyright Wiley, Chichester, 2004.

Another way to think of the depletion forces is as a result of a local increase in the osmotic pressure. When two bodies in a solvent become sufficiently such that their excluded regions overlap, solvent molecules are expelled from the area between these two bodies. Such area gives rise to an osmotic pressure further pushing the two bodies together as shown in Figure 5.5.

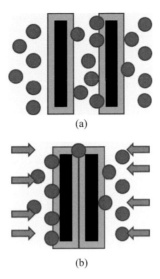

(a)

(b)

**Figure 5.5** Schematic presentation for the development of osmotic pressure as the two excluded regions of bodies overlap. (a) $h > d$, no osmotic pressure, (b) $h \leq d$ osmotic pressure is created. Note that $h$ is the distance between the two bodies, and $d$ is the diameter of solvent molecules.

Assuming a hard-sphere interaction, the osmotic pressure $(p_0)$ can be calculated as;

$$p_0 = \rho k_B T \tag{5.5}$$

Here, $\rho$ is the number density of the solvent molecules (molecule/ $m^3$).

**Example 5.2:** Calculate the depletion force between $C_{60}$ spheres in toluene at a concentration of 1.4 g/l and 300 K. Molar volume of toluene is 105.7 $cm^3$, and that of $C_{60}$ is 129 $cm^3$. Molecular weight of $C_{60}$ is 720 g/mol.

**Solution:** In order to solve this problem using Eq. 5.3 or 5.4, we need to calculate the following first; $N$, $V$, $D$, $d$, and $h$.

Starting with 1 liter ($V = 1000$ $cm^3$) as the system volume, it should contain 1.4 g of $C_{60}$.

Number of $C_{60}$ moles = 1.4 (g)/720 (g/mol) = $19.4 \times 10^{-4}$ mole.

The number of $C_{60}$ molecules = $19.4 \times 10^{-4}$ (mole) $\times 6.023 \times 10^{23}$ (molecule/mole)

Then, the number of $C_{60}$ molecules = $116.85 \times 10^{19}$ molecules.

The volume of a $C_{60}$ molecule = 129 $(cm^3/mole)/6.023 \times 10^{23}$ (molecule/mole) = $21.42 \times 10^{-23}$ $cm^3$.

Volume of $C_{60}$ in the system = $19.4 \times 10^{-4}$ (mole) $\times$ 129 $(cm^3/mol)$ = 0.25 $cm^3$.

This makes the volume of toluene in the system = 1000 − 0.25 = 999.75 $cm^3$.

Then, the number of toluene moles = 999.75 $(cm^3)/105.7$ $(cm^3/mole)$ = 9.46 mole.

Then, the number of toluene molecules $(N)$ = 9.46 (mole) $\times$ 6.023 $\times 10^{23}$ = $56.97 \times 10^{23}$ molecules.

The diameter of $C_{60}$ molecule (D) is known to be 1 nm = $1 \times 10^{-7}$ cm.

The volume of a toluene molecule can be calculated as

$$105.7 \ (cm^3/mole)/6.023 \times 10^{23} \ (molecule/mole)$$
$$= 17.55 \times 10^{-23} \ (cm^3/molecule)$$

Assuming a spherical shape (this is just an assumption since we know that toluene is a flat molecule), then, the diameter of a toluene molecule $(d)$ can be calculated as

$$17.55 \times 10^{-23} = \frac{4}{3}\pi \left(\frac{d}{2}\right)^3$$ resulting in a $d$ = $3.47 \times 10^{-8}$ cm (compare

to 3.35 Å, the thickness of the molecule).

So far, we have calculated all the parameters we need except $h$ the center to center distance between the $C_{60}$ molecules. We calculated earlier that the number of $C_{60}$ molecules = $116.85 \times 10^{19}$ molecules.

Then, the number of toluene molecules

$N$ = 9.46 (mole) $\times$ 6.023 $\times 10^{23}$ = $56.97 \times 10^{23}$ molecules.

For homogeneous solution, each $C_{60}$ molecule will be surrounded by $56.97 \times 10^{23}/116.85 \times 10^{19}$ = 4875 molecules of toluene. If we arrange such canonical in a cube, the cube will have a volume = $17.55 \times 10^{-23} \times 4875 + 21.42 \times 10^{-23}$ = $85.577 \times 10^{-20}$ $cm^3$.

The edge length of such cube = $9.49 \times 10^{-7}$ cm, which will be the value of $h$.

It is clear that at such concentration (1.4 g/l), $h > (d + D)$. Hence the depletion force will be zero.

It is important to note that in this situation, the depletion force will not cause the $C_{60}$ molecules to agglomerate.

**Example 5.3:** Calculate the configurational entropy and free energy of mixing for the system described in Example 5.2. Assume ideal mixing conditions.

**Solution:** From our previous calculations it was found that in a 1000 cm$^3$, C$_{60}$ volume = 0.25 cm$^3$, and toluene volume = 999.75 cm$^3$

Volume fraction of C$_{60}$ ($X_{C60}$) = 0.25/1000 = 25 × 10$^{-5}$
Volume fraction of toluene ($X_{tol.}$) = 999.75/1000 = 9.9975 × 10$^{-1}$
Total number of moles in the 1000 cm$^3$ of the solution = 19.4 × 10$^{-4}$ + 9.46 = 9.46194 moles.

We need to convert the units of R from J/mole·K into J/K by multiplying times the number of moles in the system

$R' = 8.314$ (J/mole·K) × 9.46194 (mole) = 78.667 (J/K)
$S_{conf.} = -78.667$ [25 × 10$^{-5}$ ln(25 × 10$^{-5}$) + 9.9975 × 10$^{-1}$ ln(9.9975 × 10$^{-1}$)] = +0.1798 (J/K) = $S^m$

Note that the configurational entropy increased due to the mixing.

Assuming ideal mixing ($H^m$ = 0) enthalpy of mixing = 0

Then, the free energy of mixing ($G^m$) = $-T \cdot S^m$ = $-300$ × 0.1798 = 53.93 J.

## 5.2 Surface Tension Effects

Surface tension effects are very important once our system is into the nanoscale. It is very well accepted that any solid body with density higher than that of a liquid cannot float on the liquid and must sink. As shown in Figure 5.6, the body will be essentially under the effect of two forces. A downward gravitational force $w$ and an upward buoyancy force $f$.

$$w = \rho_s V g \tag{5.5}$$

$$f = \rho_f V g \tag{5.6}$$

where $V$ is the body volume, $g$ is the gravitational constant, and $\rho$ is mass density. The subscripts $s$ and $f$ refer to the solid and fluid, respectively. It is clear that for the solid body to float on the surface liquid, $f$ must be higher than $w$, hence

$$\rho_f > \rho_s \tag{5.7}$$

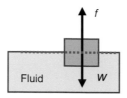

**Figure 5.6** Typical force balance considered in engineering applications.

However, there is a third upward force that we typically ignore in our engineering practice at the bulk scale that we typically use. This force is the surface tension force ($T$) as shown in Figure 5.7.

$$T = \sigma_f * l_{sf}\sin \theta \qquad (5.8)$$

where $\sigma_f$ is the surface tension of the fluid, $\theta$ is the contact angle, and $l_{sf}$ is the length of the solid/fluid interface.

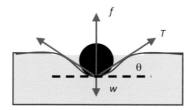

**Figure 5.7** More accurate force balance that needs to be considered in nano-scale engineering.

The correct force balance on the solid body must account for the surface tension force, hence for the body to float, Eq. 5.9 or 5.10 must apply.

$$f + T > w \qquad (5.9)$$

$$\rho_f V g + \sigma_f l_{sf}\sin \theta > \rho_s V g \qquad (5.10)$$

Comparing equations 5.7 and 5.10, it is clear that the size ($V$), shape $l_{sf}$, fluid nature $\sigma_f$, and fluid solid interaction $\theta$ play roles in Eq. 5.10, whereas only solid and fluid densities are important in Eq. 5.7 to better understand the concept. Let's try the following example.

**Example 5.4:** Calculate the resultant force acting of a steel cube on the surface of water if the cube has 1 cm, and 1 nm edge length. Water surface tension 70 mN/m, water density 1000 kg/m³, contact angle is 85°, and steel density is 7700 kg/m³.

**Solution:** To solve this problem, we need to calculate the forces acting of each cube and add them together to calculate the resultant force. We will consider upward force as positive and downward forces as negative.

| Cube edge | Vol $(m^3)$ | $l_{sf}$ $(m)$ | $f(N)$ | $T$ $(N)$ | $W$ $(N)$ | $R$ $(N)$ |
|-----------|-------------|----------------|--------|-----------|-----------|-----------|
| 1 cm | $10^{-6}$ | $4 \times 10^{-4}$ | $+9.8 \times 10^{-3}$ | $+4.9 \times 10^{-6}$ | $-7.6 \times 10^{-2}$ | $-66 \times 10^{-3}$ |
| 1 nm | $10^{-27}$ | $4 \times 10^{-9}$ | $+9.8 \times 10^{-24}$ | $+4.9 \times 10^{-11}$ | $-7.6 \times 10^{-23}$ | $+49 \times 10^{-12}$ |

From the resultant calculation shown above, it is clear that a 1 cm steel cube will sink while a 1 nm steel cube will float on the surface of water. It is also important to realize the magnitude ratio between the weight and surface tension in both cases. On the cm length scale the weight was 4 orders of magnitude higher that the surface tension. On the nanometer length scale, however, the surface tension is 12 orders of magnitude higher than the weight.

In the previous example, we have seen how surface tension effect plays an extremely important role in the nanolength scale. It is important to note that as the length scale of the system changes different forces play the major role in system's behavior leading to different behavior.

## Problems

1. Define the depletion forces, and state their existence.
2. Calculate the depletion force between 5 μm polyethylene spheres in water at 300 K at concentration 3 g/L (consider PE relative density as 1.1).
3. Calculate the osmotic pressure for water at 300 K.
4. Calculate the resultant force acting on a steel rod with 1 mm radius and aspect ratio of 100
5. Calculate the aspect ratio $(l/r)$ for a 1 mm radius steel rod to float on water.

# Chapter 6

# Symmetry and Symmetry Operations

## 6.1   Introduction

Symmetry is such an amazing natural concept. It can be observed in almost everything around us including, not surprisingly, ourselves. Everyone knows that our right half is indistinguishable from our left half if reflected in a mirror. Nature used symmetry billions of years before man and must have taught mankind how to understand symmetry and use it. The German mathematician Hermann Weyl (1885–1955) said that through symmetry, man always tried to perceive and create order, beauty, and perfection. The correlation between symmetry and beauty would become clear if one was ever moved by the symmetry, or beauty, of a snowflake, a crystal, or a flower. Ancient Greeks used the term symmetry to describe "proportionality" or similarity in arrangements of parts. They applied such understanding to create the most beautiful sculptures for humanity. Symmetry and beauty can also be heard in music and poetry. In the broadest sense, symmetry can be defined as the opposite of chaos and randomness. This definition of symmetry drags attention to the deep relationship between symmetry and entropy—a subject that is beyond the scope of this section.

The mathematical formulation of symmetry was rigorously developed in the 19th century. According to Hermann Weyl, *an*

*Gigantic Challenges, Nano Solutions: The Science and Engineering of Nanoscale Systems*
Maher S. Amer
Copyright © 2022 Jenny Stanford Publishing Pte. Ltd.
ISBN 978-981-4877-74-9 (Hardcover), 978-1-003-14704-6 (eBook)
www.jennystanford.com

*object is called symmetrical if it can be changed "somehow" to obtain the same object.* In this section we will explain what is meant by "somehow," and discuss the mathematical description of symmetry as related to molecules and crystals, the building blocks of materials. The mathematical description of symmetry is concerned with the correspondence of positions on opposite sides of a point, a line, or a plane. Mathematicians realized that at most, five different elements of symmetry are needed to fully describe the correspondence of two point positions. In other words, one would need at most five different elements of symmetry that once operated separately on a point, the point can be moved to a new indistinguishable position. To correlate this concept more closely with molecules, one may note that every molecule would possess one or more symmetry elements that once operated the molecule will assume a new configuration indistinguishable from the original one. Hence, the "*somehow*" turns out to be simply the operation of any of the following five symmetry elements.

## 6.2 Symmetry Elements and Their Operations

For symmetry, we need to have symmetry elements and symmetry operations. A *symmetry element* is a geometrical entity around which *a symmetry operation* is performed. A symmetry element can be a point, axis, or plane. A symmetry operation is the movement of a body (molecule, crystal, shape, etc.) around such symmetry element. If after the movement the shape appears the same as before, then, the geometry possesses that symmetry element. We will start with describing the five different symmetry elements then we will distinguish, further between the symmetry elements and operations.

### 6.2.1 Identity (E)

The identity symmetry element exists in everything in the universe. It is usually given the symbol (E) for the German word '*Einheit*' [1] meaning unity. Loosely, the word can be translated as "the same" or "identical."

---

[1] In crystallography and spectroscopy, the reader will come across many German nomenclature due to the ground-breaking work done by German scientists in these fields.

## 6.2.2  Center of Symmetry ($i$)

A center of symmetry is a point in space that occupies a midpoint on a line connecting two indistinguishable positions. The center of symmetry is also known as *inversion center,* hence, the designation "$i$." If one connects a line from an atom in a molecule, or generally speaking, a site or position in space, through a center of symmetry, extending the line for the same distance should lead to an equivalent indistinguishable atom, or position. For example, the carbon atom in a $CO_2$ molecule (see Figure 6.1 for example) occupies a center of symmetry. If we consider any of the oxygen atoms, connecting a line from that oxygen to the carbon and extending the line to an equal distance will lead us to the second oxygen atom in the molecule that is indistinguishable from the one we started with. Figure 6.1 illustrates the concept of a center of symmetry. It is important to note that a center of symmetry of a molecule may, or may not be occupied by an atom. For example, both $CO_2$ and $C_2H_2$ molecules possess a center of symmetry. While that center of symmetry is occupied by a carbon atom in the case of $CO_2$, it is unoccupied in the case of ethene, and actually lies on the midpoint between the two carbon atoms. As we mentioned above, if we operate the center of symmetry operation on one of the oxygen atoms in the $CO_2$ molecules, we will move that atom to the second oxygen atom position. Now what if we operate the center of symmetry on the same oxygen atom twice? This will bring the oxygen atom back to its original (i.e., identical) position. It is clear, then, that operating a center of symmetry element twice (this is mathematically expressed as $i^2$) is equivalent to the identity element of symmetry ($E$). Mathematically this can be expressed as $i^2 \equiv E$.

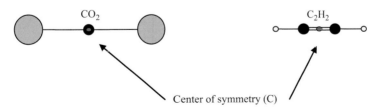

**Figure 6.1**   Illustration of the center of symmetry in two linear molecules; $CO_2$ and $C_2H_2$.

### 6.2.3 Rotation Axes ($C_n$)

If one could rotate a molecule (clockwise or anticlockwise) about an axis into a new configuration that is indistinguishable from the original configuration, the molecule is said to possess a rotation axis of symmetry ($C_n$). This symmetry element is also referred to as proper rotation axis. The subscript "$n$" is the rotation order describing the angle of rotation required to reach an indistinguishable configuration of the molecule. The rotation angle can be $2\pi/n$ ($360°/n$). For crystal, due to space filling requirements, $n$ can only assume the values 1, 2, 3, 4, and 6. For molecules, however, $n$ may take any integer value (1, 2, 3, .... ∞). For example, a $CO_2$ molecule possesses 2 two-fold rotation axes ($C_2$) normal to the molecular axis as shown in Figure 6.1. As shown in the figure, if the molecular axis is in the $x$-direction, rotation of 180° ($2\pi/2$) about either the $y$-axis or the $z$-axis will bring the molecule into an indistinguishable configuration. It is also important to note that for such linear molecules, the molecular axis represents a ($C_\infty$) rotation axis. It is also simple to observe that a benzene molecule possesses a six-fold rotation axis ($C_6$) normal to its plane. If one rotates a molecule that possesses a two-fold rotation axis 180° about that axis, this operation can be expresses as $C_2^1$. Similarly, rotation of 60° about a six-fold rotation axis is expressed as $C_6^1$. Also, a rotation of 120° about a six-fold axis is expressed as $C_6^2$. Both operations will bring the molecule into an indistinguishable configuration. A rotation of 360° about any rotation axis can be expressed as $C_n^n$. Such symmetry operation will bring the molecule into the original (identical) configuration. Hence, it is equivalent to the identity symmetry element ($E$). Mathematically this can be expressed as $C_n^n \equiv E$. Figure 6.2 illustrates the rotation axes symmetry elements possessed by a $AB_4$ type planar molecule.

### 6.2.4 Planes of Symmetry ($\sigma$) (Mirror Planes)

If a molecular configuration can be divided by a plane into two parts that are mirror image of each other, then the molecule possesses a symmetry plane ($\sigma$). This symmetry element is also known as *mirror plane*. If the molecule possesses two symmetry planes intersecting

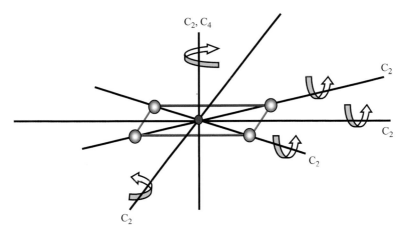

**Figure 6.2** Illustration of the rotation axes symmetry elements possessed by an $AB_4$ planar molecule.

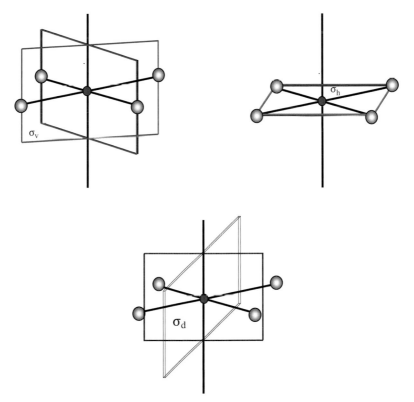

**Figure 6.3** Different types of symmetry planes possessed by an $AB_4$ planar molecule.

in a line, the intersection line will be a rotation axis. Three different types of symmetry planes have been distinguished: vertical ($\sigma_v$), horizontal ($\sigma_h$), and diagonal ($\sigma_d$) symmetry planes. $\sigma_v$ is used to denote symmetry planes intersecting in a rotation axis. If the symmetry planes are bisecting the angle between two successive two-fold axes, the planes are denoted as diagonal planes ($\sigma_d$). Horizontal symmetry planes ($\sigma_h$) are usually those in the plane of planar molecules. Some texts use primes to distinguish different types of symmetry planes. Usually, symmetry planes in a plane of planar molecules are designated as ($\sigma$). Symmetry planes out of the molecular plane are designated as ($\sigma'$). Figure 6.3 illustrates the three different types of symmetry planes possessed by a planar molecule of the type $AB_4$. It is also important to note that reflection in a symmetry plane twice ($\sigma^2$) results in the original configuration, hence $\sigma^2 = E$.

### 6.2.5 Rotation Reflection Axes ($S_n$) (Improper Rotation)

For some molecules, an indistinguishable configuration can be reproduced by rotation a certain degree ($360°/n$) about an axis and then reflection through a reflection plane that is perpendicular to the rotation axis. Such symmetry element is denoted as a rotation reflection axis, or improper rotation axis ($S_n$). The symbol S is, again, from the word *Spiegel* meaning a mirror in German. $n$ is the order of the axis. If we consider an ethane molecule ($H_3–C–C–H_3$), the positions of any two of the indistinguishable hydrogen atoms can be exchanged by rotating 60 degrees about a vertical axis, then reflecting in a horizontal plane as shown in Figure 6.4. Such a symmetry element is denoted as ($S_6$). It is important to note that neither a $C_6$ nor a $\sigma_h$ are present on their own. Improper rotation axis can also be operated in stages exactly like rotation axes. The operation described in Figure 6.4a is expressed as $S_6^1$. Repeating the operation again results in a $S_6^2$. Figure 6.4b shows that an $S_6^2$ is indeed equivalent to a $C_3$ rotation. One can clearly understand this by realizing that in a $S_6^2$, we rotate a total of 120° (equivalent to a $C_3$) and reflect in the same horizontal plane twice ($\sigma^2 \equiv E$). Improper rotation axes also have the characteristic that $S_n^n \equiv E$, if $n$ is even. Another unique characteristic of improper rotation axis is that $S_n^{n/2} \equiv i$, if $n$ is even and $n/2$ is odd ($S_6^3$ for example).

Understanding such characteristics of symmetry elements enables easier determination of all symmetry elements of a molecule. For example, if a molecule possesses a $S_6$ symmetry element, it must have an inversion center (i). The opposite is not necessarily true.

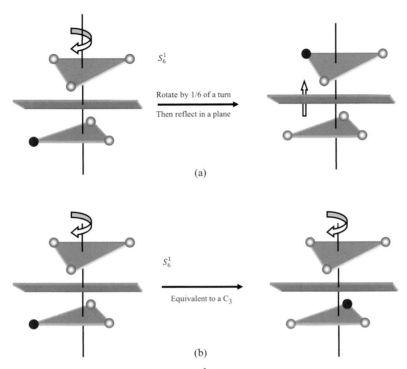

**Figure 6.4** (a) Improper reflection ($S_6^1$) in an ethane molecule. Carbon atoms are not shown for clarity. One of the hydrogen atoms was made distinguishable for illustration purpose. (b) Illustration of the equivalency of improper rotation $S_6^2$ to a $C_3$ symmetry element.

## 6.3 Symmetry Elements and Symmetry Operations

It is time to distinguish between two important concepts; symmetry elements and symmetry operations. A symmetry element is a geometrical property that could be possessed by a molecule based on the molecule exact shape. A symmetry operation is the movement of applying an action (based on a symmetry element) that results in an indistinguishable configuration of the molecule. There are only five different symmetry elements but more than five symmetry operations. For example, a six-fold improper rotation axis ($S_6$) is a symmetry element that is possessed by benzene. Such symmetry element generates different symmetry operations. For the first instant, one would say that the generated symmetry operations are six in the form of $S_6^j$ where $j$ can take any value between 1 and 6. This answer is not very accurate since, as we discussed before; the operation $S_6^2$ is equivalent to $C_3$; the operation $S_6^3$ is equivalent to $i$, the operation $S_6^4$ is equivalent to $C_3^2$, and $S_6^6$ is equivalent to $E$. This leaves the symmetry element $S_6$ capable of generating only two unique symmetry operations, namely, $S_6^1$, and $S_6^5$. This difference between point symmetry elements and operations should be clear in mind when we try to classify molecules into point groups according to their symmetry operations.

## 6.4    Point Groups

Mathematically, a set of operations (A, B, C, etc.) forms a group if the following four rules are obeyed:

1. The product of any two members of the group and the square of any member are also members of the group.
2. There must be an identity element in the group.
3. Combination of members must be associative, i.e., (AB)C should equal A(BC).
4. Every member must have an inverse that is also a member of the group, i.e., if A is a member, then $A^{-1}$ must also be a member, realizing that $AA^{-1} = E$ (the identity operation).

Hence, sets of point symmetry operations can be grouped into groups according to their satisfactory compliant with the four rules listed above. Such groups of symmetry operations are known as "point groups." It was found that all possible point symmetry operations can be grouped into 32 point groups.

### 6.4.1    Point Groups of Molecules

Point groups are given symbols made of a capital letter and, usually, two subscripts; a number and a lowercase letter, $C_{2v}$ for example.

The number indicates the order of the principal axis of the molecule. The principal axis is taken as the highest order axis, and usually defines the vertical direction. The capital letter is "D" (for dihedral) if an $n$-fold principal axis is accompanied by $n$ two-fold axes at right angle to it; otherwise, the letter will be a "C" (for cyclic). The small letter is "h" if a horizontal plane is present. If $n$ vertical planes are present, the small letter is v for a C group, and a d for a D group. It is important to note that h takes precedence over v or d. If no vertical or horizontal planes are present, the small letter is omitted. In addition, special point group symbols are reserved for molecules with certain shapes. If the molecule has a tetrahedral shape, the group symbol is "$T_d$". The group symbol is "$O_h$" for octahedral molecules, and is "$I_h$" for molecules with dodecahedral or icosahedral shapes. Such notation is known as *Shönflies* notation rule. Table 6.1 lists all possible 32 point groups for molecules.

**Table 6.1**   The 32 point groups for molecules

| Point group | Symmetry operations | Simple description | Example |
|---|---|---|---|
| $C_1$ | E | No symmetry | lysergic acid |
| $C_s$ | E $\sigma_h$ | Planar, no other symmetry | hypochlorous acid |
| $C_i$ | E, $i$ | Inversion center | 1,2-dichloroethane |

(*Continued*)

**Table 6.1** (*Continued*)

| Point group | Symmetry operations | Simple description | Example |
|---|---|---|---|
| $C_2$ | E $C_2$ | | hydrogen peroxide |
| $C_{2h}$ | E $C_2$ $i$ $\sigma_h$ | Planar with inversion center | trans-1,2-dichloroethylene |
| $C_{2v}$ | E, $C_2$, $\sigma_h$, $\sigma_v$ | Angular or see-saw | water |
| $C_3$ | E, $C_3$ | Propeller | triphenylphosphine |
| $C_{3h}$ | E, $C_3$, $C_3^2$, $\sigma_h$, $S_3$, $S_3^5$ | | boric acid |
| $C_{3v}$ | E $2C_3$ $3\sigma_v$ | Trigonal pyramidal | ammonia |
| $C_4$ | E, $C_4$ | | None |
| $C_{4h}$ | E, $\sigma$, $C_4$, $i$ | | C4H4Cl4? |
| $C_{4v}$ | E, $2C_4$, $C_2$, $2\sigma_v$, $2\sigma_d$ | Square pyramidal | xenon oxytetrafluoride |

| Point group | Symmetry operations | Simple description | Example |
|---|---|---|---|
| $C_6$ | E, $C_6$ | | None |
| $C_{6h}$ | E, $\sigma$, $C_6$, $i$ | | None |
| $C_{6v}$ | E, $6\sigma$, $C_6$ | | None |
| $C_{\infty v}$ | E, $C_\infty$ $\sigma_v$ | Linear | H——Cl 127.4 pm hydrogen chloride |
| $D_2$ | E, $3C_2$ | | None |
| $D_{2h}$ | E, 3 $\sigma$, $3C_2$, $i$ | Planar with inversion center | ethylene |
| $D_{2d}$ | E, $2S_4$, $3C_2$, $2\sigma_d$ | 90° twist | allene |
| $D_3$ | E, $C_3$, $3C_2$ | Triple helix | Tris(ethylenediamine) cobalt(III) cation |
| $D_{3h}$ | E, $2C_3$, $3C_2$, $\sigma_h$ $2S_3$ $3\sigma_v$ | Trigonal planar or trigonal bipyramidal | boron trifluoride |
| $D_{3d}$ | E $2C_3$ $3C_2$ $i$ $2S_6$ $3\sigma_d$ | 60° twist | cyclohexane |

(*Continued*)

**Table 6.1**    (*Continued*)

| Point group | Symmetry operations | Simple description | Example |
|---|---|---|---|
| $D_4$ | E, $C_4$, $4C_2$ | | cyclobutane |
| $D_{4h}$ | E $2C_4$, $C_2$ $2C_2'$ $2C_2$ $i$ $2S_4$ $\sigma_h$ $2\sigma_v$ $2\sigma_d$ | Square planar | xenon tetrafluoride |
| $D_{4d}$ | E $2S_8$ $2C_4$ $2S_8{}^3$ $C_2$ $4C_2'$ $4\sigma_d$ | 45° twist | dimanganese decacarbonyl |
| $D_{5h}$ | E $2C_5$ $2C_5{}^2$ $5C_2$ $\sigma_h$ $2S_5$ $2S_5{}^3$ $5\sigma_v$ | Pentagonal | $C_{70}$ fullerene |
| $D_{5d}$ | E $2C_5$ $2C_5{}^2$ $5C_2$ $i$ $3S_{10}{}^3$ $2S_{10}$ $5\sigma_d$ | 36° twist | ferrocene (staggered rotamer) |

| Point group | Symmetry operations | Simple description | Example |
|---|---|---|---|
| $D_{6h}$ | E $2C_6$ $2C_3$ $C_2$ $3C_2'$ $3C_2$ $i$ $3S_3$ $2S_6^3$ $\sigma_h$ $3\sigma_d$ $3\sigma_v$ | Hexagonal | benzene |
| $D_{\infty h}$ | E, $C_\omega$ $\infty\sigma$, $\infty C_2 i$, $S_\infty$ | Linear with inversion center | $O=C=O$ 116.3 pm carbon dioxide |
| $T_d$ | E $8C_3$ $3C_2$ $6S_4$ $6\sigma_d$ | Tetrahedral | 108.70 pm methane |
| $O_h$ | E, $8C_3$, $9C_2$, $6C_4$, $i$ $6S_4$ $8S_6$, $3\sigma_h$, $6\sigma_d$ | Octahedral or cubic | cubane |
| $I_h$ | E $12C_5$ $12C_5^2$ $20C_3$ $15C_2$ $i$ $12S_{10}$ $12S_{10}^3$ $20S_6$ $15\sigma$ | Icosahedral | $C_{60}$ |

Assigning a molecule to certain a point group describing its symmetry is the first and most important step in understanding and predicting the molecule spectroscopic response. In 1956, Zeldin proposed a systematic method to assign molecules to their point groups. Zeldin's method does not require the listing of all symmetry operations of the molecule. However, it focuses on major symmetry elements possessed by the molecule. Figure 6.5 shows a flowchart scheme usually used to assign molecules to point groups. It is important to note that for some of the possible molecular point groups, no known molecules are listed as example. This should not

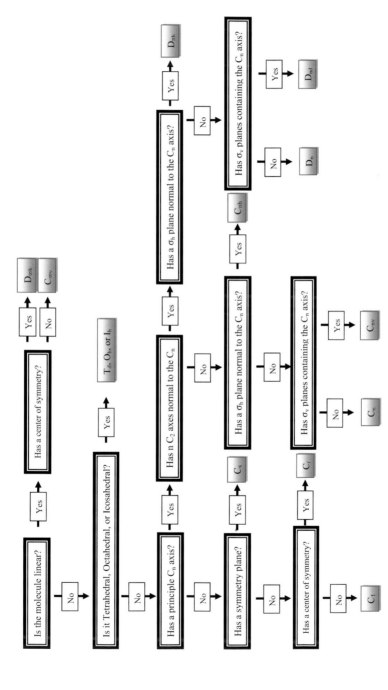

**Figure 6.5** Flow chart for assigning molecules to point groups according to symmetry elements.

be a source of confusion regarding the validity of these point groups. For example, while it has been generally believed that no molecules assume the icosahedral $(I_h)$ symmetry, three molecules have been recently known to defy such belief; borohydride anion $(B_{12}H_{12})^{2-}$, dodecahedrane $(C_{20}H_{20})$, and the molecule that was the start of nanoscale engineering [60], fullerene or $(C_{60})$.

## 6.4.2 Point Groups of Crystals

In order to consider spectroscopic investigation of solid-state materials, it is important to shed some light on the distinction between point groups as applied to molecules and point groups as applied to crystals. The difference basically comes from the crystallographic restriction theorem. The theorem is based on the observation that due to space filling restriction in solid crystals, the order of rotation axes $(n)$ should be restricted to only four values; 2, 3, 4, and 6. With such restriction in mind, applying the five point symmetry elements on the seven possible crystal lattices (triclinic, monoclinic, orthorhombic, tetragonal, rhombohedral, hexagonal, and cubic), results, again, in 32 point groups that are referred to as crystallographic point groups or crystal classes. Table 6.2 lists the 32 crystallographic point groups using *Shönflies notation*. Another notation that is also used to denote point groups is known as *Hermann–Mauguin notation*. It is named after the German crystallographer Carl Hermann and the French mineralogist Charles Mauguin. This notation is sometimes called *international notation* and used mostly in crystallography. Table 6.2 also lists the corresponding Hermann–Mauguin notation for the 32 crystallographic point groups. A set of graphical symbols was also developed for symmetry elements. While not widely used in spectroscopy, the graphical symbol system is widely used in crystallography for stereographic representation of the 32 point groups. Table 6.3 shows the graphical symbols for the different symmetry elements and their corresponding *international* notation. For the sake of completeness, Table 6.4 shows the stereographic presentation of the 32 crystallographic point groups and their corresponding international notation as correlated to the seven crystal systems.

**Table 6.2** The 32 crystallographic point groups, corresponding international notations, and related crystal systems

| Point Group Shönflies Notation | Symmetry Operations | Hermann–Mauguin Notation | Related Crystal System |
|---|---|---|---|
| $C_1$ | E | 1 | Triclinic |
| $C_s$ | E $\sigma_h$ | m | Monoclinic |
| $C_i$ | E, $i$ | $\bar{1}$ | Triclinic |
| $C_2$ | E $C_2$ | 2 | Monoclinic |
| $C_{2h}$ | E $C_2$ $i$ $\sigma_h$ | 2/m | Monoclinic |
| $C_{2v}$ | E, $C_2$, $\sigma_h$, $\sigma_v$ | mm2 | Orthorhombic |
| $C_3$ | E, $C_3$ | 3 | Rhombohedral |
| $C_{3h}$ | E, $C_3$, $C_3{}^2$, $\sigma_h$, $S_3$, $S_3{}^5$ | $\bar{6}$ | Hexagonal |
| $C_{3v}$ | E $2C_3$ $3\sigma_v$ | 3m | Rhombohedral |
| $C_4$ | E, $C_4$ | 4 | Tetragonal |
| $C_{4h}$ | E, $\sigma$, $C_4$, $i$ | 4/m | Tetragonal |
| $C_{4v}$ | E, $2C_4$, $C_2$, $2\sigma_v$, $2\sigma_d$ | 4mm | Tetragonal |
| $C_6$ | E, $C_6$ | 6 | Hexagonal |
| $C_{6h}$ | E, $\sigma$, $C_6$, $i$ | 6/m | Hexagonal |
| $C_{6v}$ | E, $6\sigma$, $C_6$ | 6mm | Hexagonal |
| $D_2$ | E, $3C_2$ | 222 | Orthorhombic |
| $D_{2h}$ | E, $3\sigma$, $3C_2$, $i$ | mmm | Orthorhombic |
| $D_{2v}$ | E, $2S_4$, $3C_2$, $2\sigma_d$ | $\bar{4}2m$ | Tetragonal |
| $D_3$ | E, $C_3$, $3C_2$ | 32 | Rhombohedral |
| $D_{3h}$ | E, $2C_3$, $3C_2$, $\sigma_h$ $2S_3$ $3\sigma_v$ | $\bar{6}m2$ | Hexagonal |
| $D_{3v}$ | E $2C_3$ $3C_2$ $i$ $2S_6$ $3\sigma_d$ | $\bar{3}m$ | Rhombohedral |
| $D_4$ | E, $C_4$, $4C_2$ | 422 | Tetragonal |
| $D_{4h}$ | E, $2C_4$, $5C_2$, $i$ $2S_4$, $\sigma_h$ $2\sigma_v$ $2\sigma_d$ | 4/mmm | Tetragonal |
| $D_6$ | E, $C_6$, $6C_2$, $6\sigma_d$ | 622 | Hexagonal |
| $D_{6h}$ | E, $2C_6$ $2C_3$ $6C_2$, $i$ $3S_3$ $2S_6{}^3$ $\sigma_h$, $6\sigma_d$ | 6/mmm | Hexagonal |

| Point Group Shönflies Notation | Symmetry Operations | Hermann–Mauguin Notation | Related Crystal System |
|---|---|---|---|
| S4 | E, $S_4$, $C_2$ | $\bar{4}$ | Tetragonal |
| S6 | E, $S_3$, $i$ | $\bar{3}$ | Rhombohedral |
| T | E, $4C_3$, $3C_2$, | 23 | Cubic |
| $T_h$ | E, $3\sigma$, $4C_3$, $3C_2$, $i$ | $m\bar{3}$ | Cubic |
| $T_d$ | E, $6\sigma$, $4C_3$, $3_c2$ | $\bar{4}3m$ | Cubic |
| O | E, $3C_4$, $4C_3$, $6C_2$ | 432 | Cubic |
| $O_h$ | E, $9\sigma$, $3C_4$, $4C_3$, $3C_2$, $i$ | $m\bar{3}m$ | Cubic |

**Table 6.3** Symbols for the symmetry elements used in stereographic representation of the 32 point groups

| Symmetry Element | Symbol in Stereogram | International Symbol |
|---|---|---|
| $E(\equiv C_1)$ | none | 1 |
| $i\ (\equiv S_1)$ | none | $\bar{1}$ |
| $\sigma$ | solid line or bold circle | m |
| $C_2$ | | 2 |
| $C_3$ | | 3 |
| $C_4$ | | 4 |
| $C_6$ | | 6 |
| $S_2$ | as for mirror plane | $\bar{2}(\equiv m)$ |
| $S_3$ | | $\bar{3}$ |
| $S_4$ | | $\bar{4}$ |
| $S_6$ | | $\bar{6}(\equiv 3/m)$ |

**Table 6.4** The 32 crystallographic point groups in stereographic representation

| Crystal system | Point groups |
|---|---|

Triclinic: $1$, $\overline{1}$

Monoclinic: $2$, $m$, $2/m$

Orthorhombic: $222$, $mm2$, $mmm$

Tetragonal: $mmm$, $\overline{4}$, $4/m$, $422$, $4mm$, $\overline{4}2m$, $4/mmm$

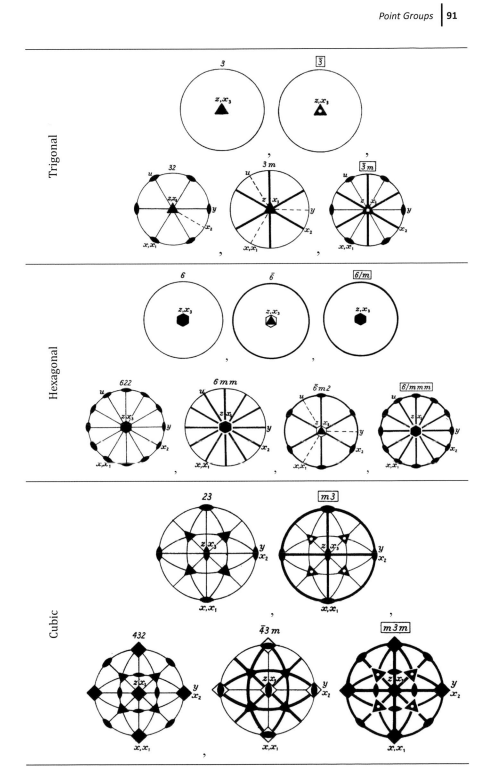

## 6.5  Space Groups

If one raises the question "why the previous 32 groups were called *point* groups?" the answer will be: because they are based on five symmetry elements each of which requires a fixed *point* during its operation. Operations about a proper or improper rotation axis requires that the axis is at a fixed point, reflection through a plane requires that the plane position is fixed at a point, a center of symmetry, itself, is a fixed point. This answer leads, logically, to two more questions: is there symmetry elements that do not require a fixed point for operation? and would operation of such symmetry elements result in other symmetry groups? The answer for both questions is "yes." This leads us to discuss two new space symmetry elements and how they produce 230 space groups once three-dimensional (3-D) solid state is considered.

Before discussing the two additional symmetry elements related to space symmetry, we should emphasize the translation operation. The translation operation along any of the unit cell principal axis is an operation to build 3-D crystals by repeating their building blocks (the unit cell) along the three principal axes of the crystal or space. From this viewpoint, translation operation is not exactly a symmetry operation. However, from a pure symmetry viewpoint, translation operation can still be considered equivalent to the identity symmetry operation.

### 6.5.1  Screw Axis (*np*)

A screw axis (also known as helical axis or twist axis) is a symmetry element that involves rotation about an axis followed by a translation along the same axis. *n* the rotation order, can be any allowed value according to the aforementioned crystallographic restriction theorem (i.e., 2, 3, 4, and 6). Translation along the screw axis is measured as a fraction of the unit cell. Such fraction takes the value $p/n$, where $p$ can assume any integer 1, 2, 3, ... $(n-1)$. For example, $2_1$ describes a screw axis symmetry operation in which an indistinguishable conformation (or an equivalent lattice site) is reached by rotating 180° about an axis then translating along that axis a distance that is half the unit cell. Similarly, a $4_3$ describes a screw axis symmetry operation in which an indistinguishable conformation (or an equivalent lattice site) is reached by rotating 90° about an axis then translating along that axis a distance that is three fourth the unit cell in that direction. Figure 6.6 demonstrates the operation of a two-fold screw axis $2_1$. Based upon the aforementioned restrictions on both $n$ and $p$ values, it is clear that only one two-fold screw axis is possible ($2_1$). Similarly, only two three-fold screw axes are possible ($3_1$ and $3_2$). Interestingly enough, objects possessing a $3_1$ screw axis and those possessing a $3_2$ screw axis are mirror image (*enantiomorphous*) of each other. The same applies to $4_1$ and $4_3$, $6_1$ and $6_5$ as well as $6_2$ and $6_4$.

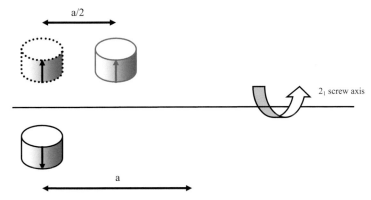

**Figure 6.6**    The operation of a two-fold screw axis $2_1$.

## 6.5.2    Glide Planes

A glide plane symmetry element involves reflection across a plane of symmetry followed by a translation parallel to that plane. Glide planes are usually denoted as a, b, or c depending on which of the principal crystal axis the glide plane is along. If the glide is along a face-diagonal direction it is denoted $n$, and if it is along a body-diagonal direction it is denoted $d$. Figure 6.7 shows the operation of a glide plane symmetry element. Comparing the resulting configurations in Figures 6.6 and 6.7 should reveal the difference between a $2_1$ screw axis and a glide plane. It should be noted that the two operations result in an object in the exact place but facing opposite directions.

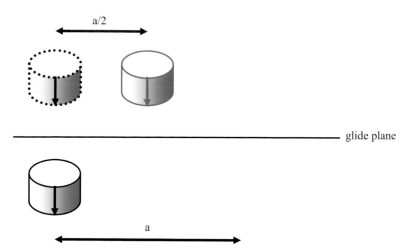

**Figure 6.7** The operation of a glide plane symmetry element.

As mathematical grouping of symmetry operations—resulting from the five point-symmetry elements—resulted in 32-point groups. Grouping of symmetry operations resulting from the seven symmetry elements—five-point elements plus two space elements—also results in 230 symmetry groups in a 3-D space. These groups are known as *space Groups.* Space groups are most commonly denoted using the Hermann–Mauguin notations. Tables (similar to Table 6.2) do exist to convert this notation to Shönflies notation. A Hermann–Mauguin notation for a space group consists of a set of four symbols. The first is a letter that describes the Bravais centering of the lattice (i.e., primitive, body centered, face centered, etc.). Table 6.5 lists the symbols used to describe the different types of cells. The other three symbols describe the most prominent symmetry elements possessed by the crystal. The first of these symbols describes the symmetry of the principal axis of the cell. The next symbol describes the next major symmetry element, etc. For example, $P2_1/m$ indicates a primitive cell with a $2_1$ screw axis as the principal axis and a mirror plane perpendicular to the principal axis. Also, $Pmc2_1$ indicates a primitive cell with a mirror plane, a glide plane and a $2_1$ screw axis.

**Table 6.5** Symbols for the six possible Bravais lattice centerings

| Type | Symbol | Example |
|---|---|---|
| Primitive (lattice sites on corners only) | P | $a \neq b \neq c$ |
| Face Centered (lattice sites on corners and all face centers) | F | $a \neq b \neq c$ |
| Side Centered (lattice sites on corners and centers of two of the faces) | A, B, or C | C centering is shown; $a \neq b \neq c$ |
| Body Centered (lattice sites on corners and body center) | I | |
| Rhombohedral | R | $\alpha, \beta, \gamma \neq 90°$ |

Primitive centering (P): lattice points on the cell corners only
Body centered (I): one additional lattice point at the center of the cell
Face centered (F): one additional lattice point at center of each of the faces of the cell
Centered on a single face (A, B or C centering): one additional lattice point at the center of one of the cell faces

## 6.6    Space Groups in 1- and 2-D Space

Realizing that space groups describe the symmetry of objects within space, and that they were developed based upon a number of symmetry operations available in that space, leads to the conclusion that the possible space groups depend on the dimensionality of the space in which they exist. Since, traditionally, materials scientists are inherently interested in 3-D space representing materials system, we have focused our discussion on 3-D space groups. It is important, however, to know that 1- and 2-D space groups have also been investigated and are of great interest and applications in other fields such as theoretical physics, art, architecture, and textile industry. In materials science, 1- and 2-D liquid crystals as well as symmetry along a single polymer or a co-polymer molecule is a subject of increasing interest. In addition, as nanotechnology brings quasi 1-D (nanowires, nanotubes, one-dimensional crystals, etc.) and 2-D (graphene sheets, self-assembled layers, etc.) material systems into attention, a discussion of space groups in these spaces will definitely be helpful if properties of 1- and 2-D materials are to be well understood.

### 6.6.1    Space Groups in 1-D Space (Linear Objects)

Considering 1-D spaces, spatial restrictions make only certain symmetry elements and operations possible. In addition to translation along the one dimension of the space, rotation about an axis normal to the line is possible as long as the rotation order is restricted to 180°(two-fold rotation axes only). We may also realize that inversion through a point on the line, reflection through a plane normal to the line, and glide through a plane passing through the line are also allowed. These allowed symmetry operations in one dimensional space result in *seven different 1-D space groups* also known as *Frieze Groups*. These groups can belong to only two types, one type with reflection and a second type with no reflection.

Symbols for 1-D space groups are made of four characters. The symbol usually starts with "F" letter (for frieze). In certain texts, the letter P (for primitive cells) is also used. The second character is a number which is 2 if rotational symmetry does exist, and is 1 if rotational symmetry does not exist. The third character is *m* if there is a transverse mirror, and is 1 if no transverse mirror exists. The fourth character is *m* if a longitudinal mirror exists, and is a *g* if a longitudinal glide exists. The fourth character is omitted if no longitudinal symmetry exists. Table 6.6 lists the seven possible 1-D space groups (frieze groups) and the symmetry elements possessed by each of them. It is also important to note the importance of frieze groups in describing symmetry along polymer molecules, carbon nanotubes, and whiskers.

**Example 6.1:** Determine the space group of the chain shown below.

**Solution:** It is clear that the repeating unit in the 1-D sequence above has mirror plane symmetry along the linear axis, and a transverse mirror plane normal to the linear axis. This, as we discussed in the symmetry elements requires the presence of a two-fold axis along the line of intersection between the two mirror planes as shown below. Such repeating unit is translated along the line. Hence the linear space group will be F2mm.

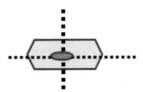

Another way to look at the problem is to consider half the elongated hexagon as the repeating unit (the shaded area) as shown below:

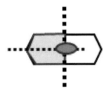

It is clear that the repeating unit still possesses a longitudinal mirror plane and is translated using a transverse mirror element. And since the presence of two intersecting planes necessitates the presence of a two-fold rotation, then the space group is still F2mm.

**Table 6.6** The seven possible one-dimensional space groups (Frieze groups)

| Frieze Group | Symmetry Elements | Example |
|---|---|---|
| F1 | Translation only | |
| F2 | Translation, 2-fold rotation | |
| F1m | Translation, Transverse mirror | |
| F11m | Translation, longitudinal mirror | |
| F2mm | Translation, rotation, transverse and longitudinal mirrors | |
| F11g | Translation, glide | |
| F2mg | Translation, rotation, transverse mirror, glide | |

### 6.6.2  Space Groups in 2-D Space (Plane Space Groups or Wallpaper Groups)

Considering the restrictions of a 2-D space, it was shown mathematically that only 17 space groups are possible in such a case. 2-D space groups are also referred to as *plane space groups* or *wallpaper* groups. The name wallpaper groups originated from the fact that these symmetry objects have been used for centuries as decorative art for textiles, architect, and for wallpaper designs. Understanding the group symmetry in 2-D space is crucial for the study and better understanding of the new emerging field of 2-D materials. The symbols for wallpaper groups start with a letter describing the cell centering (i.e., P for primitive or C for centered) followed by an integer describing the highest order of rotation, followed by *m, g,* or 1 to reflect the existence of a mirror plane, a glide plane, or no symmetry element, respectively. Table 6.7 demonstrates the 17 possible plane space groups and their symmetry features. Note the importance of different colors in determining the symmetry group of the geometry.

**Table 6.7**  The seventeen possible plane space groups in two-dimensional space

| Wall Paper Group | Symmetry Elements | Example |
|---|---|---|
| P1 | Translation only along any two axes in the plane | |
| Pg | Glide reflections only on parallel axes | |

*(Continued)*

**Table 6.7** (*Continued*)

| Wall Paper Group | Symmetry Elements | Example |
|---|---|---|
| Cm | Parallel reflection planes with at least one glide half way between the mirror planes | |
| Pm | Only reflection planes all parallel | |
| P6 | One six-fold rotation (60°), two three-fold rotation (120°), and three two-fold rotation (180°) no mirror or glide reflections | |
| P6mm | One six-fold rotation (60°), two three-fold rotation, and three two-fold rotation. In addition, reflections in six distinct directions plus glide reflections in six distinct directions, whose axes are located halfway between adjacent parallel reflection axes | |
| P3 | Three three-fold rotation (120°). No reflections or glide reflections | |

| Wall Paper Group | Symmetry Elements | Example |
|---|---|---|
| P3m1 | Three three-fold rotation (120°). Three reflections in the three sides of an equilateral triangle. Every rotation axis lies on a reflection axis. In addition, glide reflections in three distinct directions, whose axes are located halfway between adjacent parallel reflection planes | |
| P31m | Three three-fold rotation (120°), of which two are each other's mirror image. Three reflections in three distinct directions plus at least one rotation not on a reflection plane. Three glide reflections in three distinct directions, whose axes are located halfway between adjacent parallel reflections | |
| P4 | Two four-fold rotation axes (90°), and one two-fold rotation (180°). No reflections or glide | |
| P4m | Two four-fold rotation (90°), and reflections in four distinct directions (horizontal, vertical, and diagonals) plus glide reflections whose axes are not reflection axes; two-fold rotations (180°) are at the intersection of the glide reflection axes. All rotation lie on reflection axes. | |

(*Continued*)

**Table 6.7** (*Continued*)

| Wall Paper Group | Symmetry Elements | Example |
|---|---|---|
| P4g | Two four-fold rotation (90°), which are each other's mirror image, but it has reflections in only two directions, which are perpendicular. There are two-fold rotations (180°) whose located at the intersections of reflection axes. Also, glide reflections parallel to the reflection axes, in between them, and also at an angle of 45° with these. | |
| P2 | Four two-fold rotations (180°). No reflections or glide. | |
| Pgg | Two two-fold rotation axes, and glide reflections in two perpendicular directions. | |
| Pmg | Two two-fold rotation (180°), and reflections in only one direction. Also, glide reflections whose axes are perpendicular to the reflection axes. Rotation all lie on glide reflection axes. | |

| Wall Paper Group | Symmetry Elements | Example |
|---|---|---|
| Pmm | Reflections in two perpendicular directions, and four two-fold rotation (180°) located at the intersections of the reflection axes. | |
| Cmm | Reflections in two perpendicular directions, and three two-fold rotation (180°) of which two *are* on a reflection axis and one *is not* on reflection axis. | |

**Example 6.2:** Determine the plane space group for the geometry below.

**Solution:** Looking for symmetry elements contained in the given geometry we note that it is a primitive cell (P), its highest rotation is a six-fold rotation (6) and mirror reflections (m) in six distinct directions as shown below. This makes the geometry belong to P6mm plane group.

It is important to compare the geometry in this example to that representing the P3m1 in Table 6.7 as shown in Figure 6.8. The difference between the two geometries is the coloring of the surrounding hexagons. This difference in coloring changes the highest rotation symmetry from six-fold (in the example) to a three-fold rotation symmetry (in the P3m1).

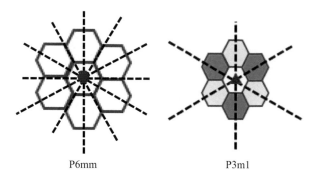

P6mm                                        P3m1

**Figure 6.8**  Effect of coloring on the planer space groups in two-dimensional space.

## 6.7 Summary

To conclude our discussions on space groups, it is important to realize that other types of space groups do exist and have been investigated. Depending on the dimensionality of the space considered different space groups evolve and become relevant. For example, in physics and mathematics when 4-D space is considered, in a time evolving systems for example, 4895 space groups are possible. In addition, magnetic space groups can be constructed from ordinary 3-D space groups by considering magnetic spins associated with lattice points. Such magnetic space groups are also known as *colored space groups*, and 1651 of them are possible. Such space groups are not within the scope of our engineering field and will not be discussed here.

### Problems

1. Discuss all the symmetry elements in a 3-D space.
2. What is the meaning of the following symmetry element and operation?
   a. $C_4^3$    b. $C_6^2$    c. $S_6^3$    d. $4_3$
3. What is the equivalent of the following symmetry operations?
   a. $\sigma^2$    b. $S_6^3$    c. $C_3^3$
4. Which point group would a water molecule belong to?

5. Which point group would a $(SiO_4)^{4-}$ molecule belong to?
6. How many space groups exist in the 1-D, 2-D, and 3-D spaces?
7. Determine the space group for the shown 2-D shapes.

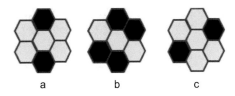

a     b     c

# Chapter 7

# Fullerenes: The Building Blocks

## 7.1   Introduction

In this chapter, fullerene as nanomaterial building blocks will be discussed. The beginning of fullerenes and their current state will be covered. From structural dimensionality viewpoint, fullerenes can be classified into zero-dimensional (0-D), 1-D, or 2-D structures. 0-D fullerenes include spheroidal cage-like nanocarbon species that were traditionally referred to as fullerenes. They include $C_{60}$, $C_{70}$, $C_{80}$, etc. 1-D fullerenes include tubular-like fullerenes including single-walled carbon nanotubes (SWCNTs), double-walled carbon nanotubes (DWCNTs), and multi-walled carbon nanotubes (MWCNTs). By 2-D fullerenes, we refer to the recently isolated graphene sheets including single-layered and multi-layered graphene sheets.

For a concept to develop into a technology, the availability of suitable building blocks is a must. Material building blocks enable experimental verification of the theory behind the concept and provide the essential ingredient enabling building devices and products bringing the new technology into reality. In our case, nanotechnology is based upon nanobuilding blocks that are essentially small thermodynamic systems, as we discussed in previous chapters. Hence, for nanotechnology, any thermodynamic small system is, indeed, a building block. Consequently, any cluster of matter, regardless of its physical size, that would be on the order of

*Gigantic Challenges, Nano Solutions: The Science and Engineering of Nanoscale Systems*
Maher S. Amer
Copyright © 2022 Jenny Stanford Publishing Pte. Ltd.
ISBN 978-981-4877-74-9 (Hardcover), 978-1-003-14704-6 (eBook)
www.jennystanford.com

any of the scale lengths discussed earlier should satisfy the definition and can be considered as a building block. To this end, many systems can be defined as building blocks. Some of these building blocks are, themselves, made of smaller building blocks such as biological cells, or even biological species. If we consider the subject from such a viewpoint, the domain of nanotechnology building blocks will be too large to consider in details. In order to keep the subject focused and of practical value, we will focus on some of the recently discovered and investigated building blocks which are essentially manmade. The word "some" in the previous statement was used intentionally since, as a matter of fact, recently developed and produced manmade nanobuilding blocks themselves are too many to discuss in details. The list is already long, and new building blocks are produced, investigated, and reported often. The reported list in the literature include nanowire, nanorods, nanocones, nanohorns, nanospheres, nanoshells, etc., each of which has been produced from a number of different materials.

Based on our previous discussion of matter in the nanodomain, it can be inferred that nature is the best designer for nanosystems. Interestingly enough, nature used one particular element most frequently in its designs, especially for living systems that is carbon. Hence, as we discuss the building blocks of nature's preferred technology, carbon-based building blocks should be a good choice for discussion. Previously in this book, we discussed nanophenomena that were observed in systems based mainly on metallic clusters in the nanodomain. In the following chapters we will focus on carbon-based building blocks, namely, fullerenes. Fullerenes are a recently discovered form of nanocrystalline carbon. They come in different geometrical shapes. Spherical, or generally speaking, balloon-like, shapes are usually referred to as fullerenes or buckyballs. Cylindrical shapes are more popularly known as SWCNTs, and more recently, single sheet form referred to as graphene has also been produced and utilized as a building block for nanotechnology. Figure 7.1 shows the different types of carbon nanospecies to be discussed in the following chapters, namely, fullerenes, SWCNTs, and MWCNTs, carbon nano-onions, and graphene sheets. In addition, examples of recently created nanohybrid structures will be demonstrated and discussed from a materials science viewpoint. Figure 7.2 shows examples of such nanohybrid structures (or sometimes referred to as mesostructures). In these structures different nanobuilding blocks, are incorporated into new structures. For example, Figure 7.2a shows a mesostructure knows as *peapod*. In this structure fullerene molecules are inserted inside SWCNTs ($C_{60}$@SWCNTs). Figure 7.2b shows a different type of structures usually referred to as *nanobuds*. In nanobuds, fullerene molecules are attached to the surface of SWCNTs. Figure 7.2c shows an example of more complex mesostructures in which several fullerene molecules are attached at different sites of the surface of SWCNTs.

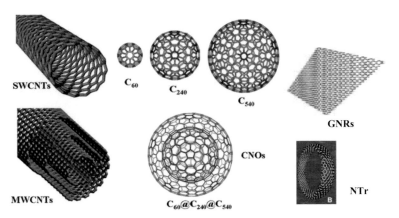

**Figure 7.1** Different types of carbon nanospecies. Single-walled carbon nanotubes (SWCNTs), multi-walled carbon nanotubes (MWCNTs), [60], [240], and [540] fullerene, carbon nano-onions (CNOs), graphene nano-ribbons (GNRs), and nanotorus (NTr). Adapted with kind permission from Delgado et al., *J. Mat. Chem.*, 2008, 18, 1417, Copyright The Royal Society of Chemistry, 2008, and from Liu et al., *Nature*, 1997, 385, 780, Copyright Nature Publishing Group, 1997.

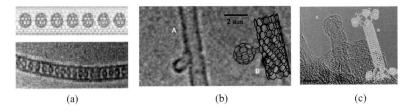

(a)                        (b)                        (c)

**Figure 7.2** Transmission electron microscopy photographs and computer-generated images of (a) peapods ($C_{60}$@SWCNT) (Reproduced with kind permission from Okazaki et al., *Physica B: Condens. Matter.*, 2002, 323 (1), 97–99. Copyright Elsevier, 2002), (b) nanobud, and (c) a mesostructure incorporating a nanotube and a number of [60] fullerenes (Reproduced with kind permission from Nasibulin et al., *Nat. Nanotechnol.*, 2007, 2, 156. Copyright Nature Publishing Group, 2007).

## 7.2 Fullerenes: The Beginnings and Current State

Historically, the possibility of creating *graphite balloons* similar to geodesic cages was first discussed in 1966 by David Jones who was writing under a pseudonym "Daedalus" in the journal the *New Scientist*. The most famous form of fullerene molecules, however, is the [60] fullerene or $C_{60}$ commonly known as buckyball. The molecule is made of 60 carbon atoms in a spherical shape that exactly resembles a soccer ball. The earliest record of a fullerene molecule, however, was in an article (in Japanese) by Eiji Osawa in 1970. In this article, Osawa speculated that such a molecule would be stable. In the following year (1971) Osawa and Yoshida described the speculated molecule in more details in a book—also in Japanese—on aromatic chemistry. Two years later, in 1973, Bochvar and Gal'pern used Hückel calculation to determine the energy levels and molecular orbitals in the $C_{60}$ molecule. Later, in 1981, Davidson applied general group theory techniques to a range of highly symmetrical molecules, one of which was the $C_{60}$. Hence, it is clear that by 1981, the idea of stable fullerene molecules did, in fact, exist and early studies have been done to explore the energy levels and molecular orbitals as well as symmetry properties of such molecule. The molecule itself was not yet experimentally observed.

In September 1985, such observation took place. While trying to simulate stellar nucleation conditions of cyanopolyyenes[1] using the, then recently, developed laser vaporization cluster technique by Richard Smalley and his co-workers at Rice University, USA, Curl, Kroto, Smalley, and their co-workers vaporized graphite and a serendipitous discovery was made. The $C_{60}$ molecule was observed and found to be remarkably stable. The molecule was named Buckminsterfullerene (later fullerene for short) after the famous architect Buckminster Fuller (1895–1983) who first created the geodesic cage or dome design which the fullerene molecule resembles. Figure 7.3 shows a photograph of Buckminster Fuller and his famous geodesic dome design.

---

[1]Some carbon chain molecules known as cyanopolyyenes were earlier discovered in interstellar regions streaming out of red giant carbon stars. Cyanopolyyenes has the form $H-C\equiv C-C\equiv C-C\equiv N$ with 5, 7, 11, and up to 33 carbon atoms.

**Figure 7.3** Buckminster Fuller and his famous geodesic dome design. Courtesy of Buckminster Fuller Institute.

This interesting and long waited discovery triggered a scientific race to investigate each and every aspect of the new molecule. At first, however, progress was slow due to the fact that the amount of $C_{60}$ produced by Smalley's method was very minute. The real race of development started in 1990 with the findings of Krätschmer of the Max Planck Institute at Heidelberg, Huffman of the University of Arizona, and their co-workers who could produce $C_{60}$ molecules in macroscopic amounts using a simpler, more accessible technique than that used by Smalley. The new technique vaporized graphite using simple carbon arc in helium atmosphere. The soot deposited on the walls of the vessel, once dispersed in benzene, produced a reddish solution. Once dried, the solution produced beautiful crystals of "fullerite" which turned to be made of 90% $C_{60}$ and 10% $C_{70}$. By using Krätschmer and Huffman method, $C_{60}$ and other allotropes of fullerenes could be produced in reasonable amounts in a way accessible to may laboratories. This accelerated the fullerene investigation race and started what Curl described as "*the Dawn of Fullerenes.*" By the year 1991 fullerenes were the subject matter of 90% of the most cited papers, and the subject is still being of current scientific interest.

In 1991, Sumio Iijima of the NEC laboratories in Japan reported "the preparation of a new type of finite carbon structure consisting of needle-like tubes, produced using an arc-discharge evaporation method similar to that used for fullerene synthesis." Electron microscopy investigation revealed that the needles consist of co-axial tubes of a number of graphic sheets ranging between 2 and 50. These new molecular cylinders of graphitic sheets were called carbon nanotubes. As a matter of fact, Iijima's report on carbon nanotubes was not the first in the literature. As early as 1952,

Radushkevich and Lukyanovich[2] reported, in the Journal of Physical Chemistry of Russia, the first TEM evidence for the tubular nature of some nanosized carbon filaments. In 1974, Oberlin, Endo, and Koyama working on benzene derived carbon fibers reported that "These fibers have various external shapes and contain a hollow tube with a diameter ranging from 20 to more than 500 Å along the fiber axis." Figure 7.4 shows the transmission electron micrographs first produced by Endo and Iijima for what we currently refer to as *carbon nanotubes*.

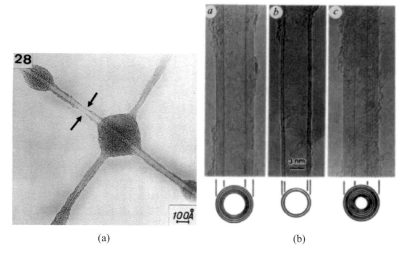

(a)                                           (b)

**Figure 7.4** (a) Electron microscopy micrograph first appeared in Endo's original Dissertation and republished in 1974. Reproduced with kind permission from Oberlin et al., *Carbon*, 1976, 14, 133. Copyright Elsevier, 1976. (b) Transmission electron micrograph and cross sections of multi-walled carbon nanotubes. Reproduced with kind permission from Iijima, *Nature*, 1991, 354, 56. Copyright Nature Publishing Group, 1991.

In 1993, Iijima and Ichihashi reported the observation of a more interesting carbon nanospecies; the SWCNTs. They found the single-shell tubes in carbon soot formed in a carbon arc chamber similar to that used for fullerene production. Figure 7.5 shows the original transmission electron micrograph of SWCNTs.

[2]Radushkevich LV, Lukyanovich VM. *O strukture ugleroda, obrazujucegosja pri termiceskom razlozenii okisi ugleroda na zeleznom kontakte. Zurn Fisic Chim*, 1952; 26:88–95.

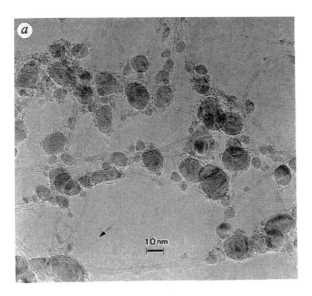

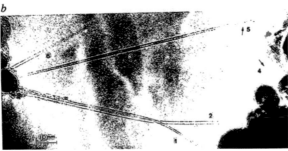

**Figure 7.5** Original electron micrograph by Iijima and Ichihashi showing single-shell carbon nanotubes. The tube labeled 1 is 0.75 nm in diameter. The tube labeled 2 is 1.37 nm in diameter. Straight and terminated (4 and 5) tubes can be seen. Reproduced with kind permission from Iijima and Ichihashi, *Nature*, 1993, 363, 603. Copyright Nature Publishing Group, 1993.

The discovery of fullerene molecules in 1985 and the later discovery of carbon nanotubes in early 1990s established a new field of carbon nanosciences and triggered an active international scientific race to investigate the structure and properties of such fascinating molecules and to discover other allotropes of this class of matter. The search for new allotropes of carbon was crowned in 2004 with the ability to isolate and manipulate single sheets of graphite currently referred to as *graphene* sheets or nanoribbons. Graphene sheets are essentially related to a much older form of graphite known as *exfoliated graphite*. In the exfoliated form, graphite is expanded by up to hundreds of times along its *c* axis. Scientific and technological developments in the field of exfoliated graphite took place in the late 1960s when flexible graphite foils were made of exfoliated graphite and used for high temperature gaskets and seals. The ability to isolate single layers of graphene, however, at the age of nanotechnology spawned intensive research into the synthesis, properties, applications, and methods of the mass production of this new form of carbon. Production methods of the graphene ribbons involve both traditional exfoliated graphite techniques and more sophisticated techniques based on unzipping of carbon nanotubes. The importance of graphene sheets lies in the fact that they provide a unique opportunity for experimental investigation of truly 2-D systems—an opportunity that scientists never had before. In their profound article, Geim and Novoselov described graphene as "the mother of all graphitic forms. It can be wrapped into a 0-D buckyballs, rolled into 1-D nanotubes, or stacked into 3-D graphite." Figure 7.6 schematically represents the different branches of the graphitic family.

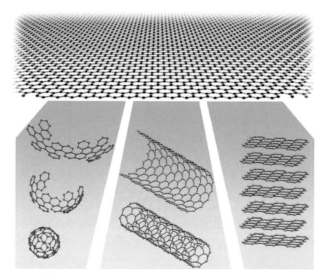

**Figure 7.6** Different branches of the graphitic family; two-dimensional graphene that can be rolled wrapped into zero dimensional fullerenes, rolled into nanotubes, or stacked into three-dimensional graphite. Reproduced with kind permission from Geim and Novoselov, *Nat. Mater.*, 2007, 6, 183. Copyright Nature Publishing Group, 2007.

More recently, carbon aromatic chains—a unique flat 1-D form of nanocarbon—were derived from graphene sheets.

In the following chapters we will discuss the structure and properties of different types of fullerenes. We can classify the fullerenes in three different classes; 0-D fullerenes that include the different types of spheroidal fullerene molecules, 1-D fullerenes that include cylindrical fullerenes such as nanotubes, and 2-D fullerenes that include flat graphene sheets and ribbons. It is important to note that while the zero-dimensionality description could be appropriate for small fullerene molecules with limited number of carbon atoms (such as $C_{60}$), it cannot be claimed accurate for giant fullerene molecules with large number of carbon atoms (such as $C_{540}$). Also, based upon our discussion of the history and beginnings of fullerene science, it is interesting to note that the dimensionality classification follows the discovery trend of these fullerenes.

## Chapter 8

# Spherical, Zero-Dimensional, Buckminsterfullerenes

## 8.1 Introduction

Spherical or zero-dimensional (0-D) fullerenes were the first type of fullerenes to be observed and investigated. In this chapter we will discuss the structure, production methods, extraction methods, and purification methods for spherical fullerenes. In addition, the mechanical, optical, and chemical properties of the mostly investigated fullerenes will be discussed. Finally, we will discuss the properties of 2-D films produced using 0-D fullerenes.

## 8.2 The Structure

Spherical or 0-D fullerene molecules are closed hollow cage-like spheroidal molecules made of a number ($n$) of covalently bonded carbon atoms. The nomenclature most commonly used is either $C_n$ or "[$n$] fullerene." The structure of the carbon cage in fullerenes consists of a number of hexagons and pentagons. It is known that carbon in its $sp^2$ hybridization state creates planar sheets of connected hexagons. The introduction of pentagons in such flat structure causes the sheet

*Gigantic Challenges, Nano Solutions: The Science and Engineering of Nanoscale Systems*
Maher S. Amer
Copyright © 2022 Jenny Stanford Publishing Pte. Ltd.
ISBN 978-981-4877-74-9 (Hardcover), 978-1-003-14704-6 (eBook)
www.jennystanford.com

to wrinkle and curve due to slight changes in the C–C bond length. According to Euler's theorem for polyhedra, the number of faces ($f$), vertices ($v$), and edges ($e$) of a polyhedron should be correlated by:

$$f + v = e + 2 \qquad (8.1)$$

Hence, if we consider a polyhedron consisting of $h$ hexagons and $p$ pentagons, the following three relations should be satisfied:

$$f = p + h, \quad 2e = 5p + 6h, \quad 3v = 5p + 6h \qquad (8.2)$$

This yields that

$$6(f + v - e) = p = 12 \qquad (8.3)$$

Then, for any spherical fullerene $C_n$ having a cluster of $n$ atoms (with $n$ always even and larger than 20), and consisting of only pentagons and hexagons, closed cage-like structures consisting of 12 pentagons and $(n-20)/2$ hexagons can be constructed. This rule is important in predicting fullerene structures. The original discovery, as we mentioned before, was for $n = 60$. Then, according to Euler's rule the structure should be a cage-like molecule made of 60 covalently bonded carbon atoms arranged in 12 pentagons and 20 hexagons.

The smallest possible fullerene molecule should be the $C_{20}$. In this case the cage structure should consist of only 12 pentagons. While, theoretical studies have noted that abutting pentagons raise the $\pi$-energy and cause structural destabilization, experimental observation of the $C_{20}$ molecule was reported in year 2000. Prinzbach et al. showed that the cage-structured $C_{20}$ can be produced from its per-hydrogenated form (dodecahedrane $C_{20}H_{20}$) by replacing the hydrogen atoms with relatively weakly bound bromine atoms, followed by gas-phase debromination. The $C_{20}$ molecules that were produced, however, were rather unstable, but their fleeting existence was confirmed using mass-selective anion photoelectron spectroscopy. Molecular simulations based on density functional theory calculations, however, confirmed the experimental observation of a [20] fullerene molecule. Other isomers of the 20 carbon cluster taking the shapes of a bowl containing both hexagons and pentagons and a ring were also reported. Figure 8.1 shows the three structural forms of a 20 carbon atoms cluster; a caged fullerene, a bowl, and a ring.

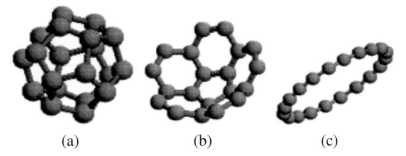

(a)            (b)            (c)

**Figure 8.1**    The three structures of a 20 carbon atoms cluster; (a) [20] fullerene, (b) a bowl shaped graphene, and (c) a ring. Reproduced with kind permission from Prinzbach et al., *Nature*, 2000, 407, 60. Copyright Nature Publishing Group, 2000.

In spite of the "pentagon rule" stating that abutting pentagons are destabilizing, $C_{20}$ molecules were produced and experimentally observed. The stability of other forms of fullerene molecules $C_n$, with $n$ = 24, 28, 32, 36, and 50 was also investigated and the structures were shown to be stable as well.

On the other extreme of the size scale for fullerene molecules, we usually find giant fullerene containing hundreds and even thousands of carbon atoms. The largest fullerene molecule reported in theoretical studies is the $C_{4860}$. It is important, however, to note that the exact shape and structure of large fullerenes is still not fully resolved. Fullerenes usually form isomers. As the number of carbon atoms in the fullerene molecule increases, the molecule can assume different geometrical shapes or structures all of which can still satisfy both the Euler and the pentagon rules mentioned above. Hence, it would be logical to realize that as the fullerene size increases the number of possible structures or isomers the molecule may assume also increases. These different possible structures of the different isomers would belong to different symmetry groups. Therefore, making definitive assignment of higher fullerene structures and symmetry becomes difficult. For example, while $C_{60}$ can be formed in only one structure that belongs to the icosahedral symmetry $(I_h)$, the $C_{80}$ can be formed in seven different structures belonging to any of six different symmetry groups; $(I_h)$, $(D_{5d})$, $(D_2)$, $(D_3)$, $(D_{5h})$, or the $(C_{2v})$ symmetry group. Figure 8.2 shows two of the seven $C_{80}$ isomers belonging to the $D_{5d}$ and the $I_h$ symmetry point groups. Table 8.1 shows the number of isomers $(N_i)$ obeying the isolated pentagon rule and symmetry point groups of common fullerene molecules.

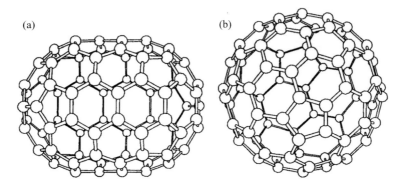

**Figure 8.2** Two isomers of the $C_{80}$ fullerene molecule belonging to (a) $D_{5d}$ point group, and (b) $I_h$ point group. Reproduced with kind permission from Schmalz et al., *Chem. Phys. Lett.*, 1986, 130, 203. Copyright Elsevier, 1986.

**Table 8.1**  Number of isomers ($N_i$) obeying the isolated pentagon rule and symmetry point groups of common fullerene molecules

| Fullerene | $N_i$ | Point Group Symmetry |
|---|---|---|
| $C_{60}$ | 1 | $I_h$ |
| $C_{70}$ | 1 | $D_{5h}$ |
| $C_{76}$ | 1 | $D_2$ |
| $C_{78}$ | 5 | $D_3$, $2D_{3h}$, $2C_{2v}$ |
| $C_{80}$ | 7 | $D_2$, $2D_{3h}$, $2C_{2v}$, $D_{5h}$, $D_{5d}$, $I_h$ |
| $C_{82}$ | 9 | $3C_2$, $C_{2v}$, $2C_{3v}$, $3C_s$ |
| $C_{84}$ | 24 | see Figure 8.3 for illustration |
| $C_{88}$ | 35 | see P. Fowler and D. Manolopoulos, *An Atlas of Fullerenes*, Clarendon Press, Oxford, 1995 for the full list |

Theoretical studies based on density functional theory (DFT) calculations show that large fullerenes of icosahedral symmetry prefer faceted over spherical shapes. Figure 8.4 shows the faceted as well as the hypothetical spherical shapes of large fullerenes. Very few of such molecules have been experimentally observed in individual forms. Many have been observed as one of the shells in a nano-onion structure. Nano-onion structures will be discussed later in a separate section.

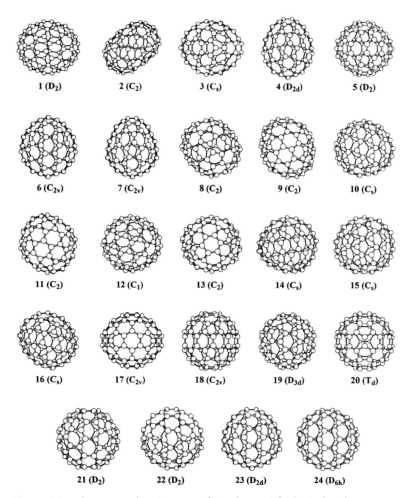

**Figure 8.3** The twenty-four isomers of $C_{84}$ that satisfy the isolated pentagon rule.

Out of the huge possible number of spherical fullerene molecules, merely five have been produced in significant quantities rendering them most accessible and making their experimental investigation attractive. These fullerene molecules are $C_{60}$, $C_{70}$, $C_{76}$, $C_{78}$, and $C_{84}$. From structural symmetry viewpoint, their structures have been identified to belong to eight different symmetry groups as follows; The $C_{60}$ molecule belongs to the $I_h$ symmetry. The $C_{70}$ belongs to the $D_{5h}$ symmetry. The $C_{76}$ molecule belongs to the $D_2$ symmetry. The

$C_{78}$ molecule has five isomers that can belong to any of the three symmetry groups $D_3$, $D_{3h}$, or $C_{2v}$. The $C_{84}$ molecule, however, has 24 isomers that can belong to different symmetry groups. Figure 8.5 shows the eight possible lowest energy structures assumed by the five most accessible fullerenes. Atoms numbering is based on the IUPAC system.

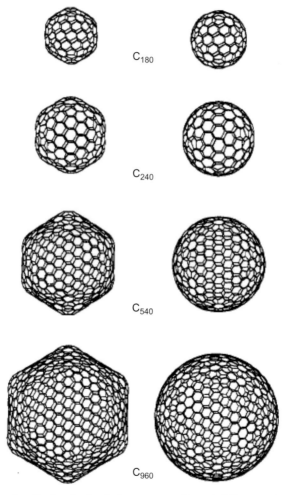

$C_{180}$

$C_{240}$

$C_{540}$

$C_{960}$

**Figure 8.4** Facetted and spherical structures of large $I_h$ fullerenes. Reproduced with kind permission from Bakowies et al., *J. Am. Chem. Soc.*, 1995, 117, 10113 Copyright American Chemical Society, 1995.

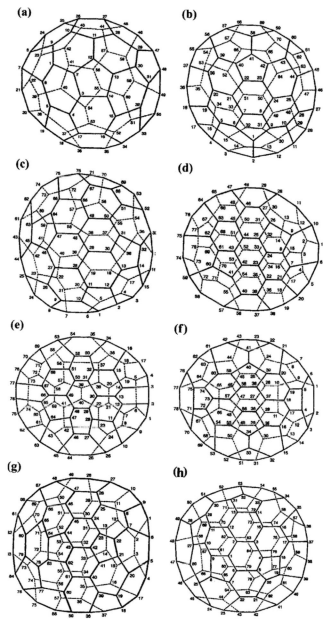

**Figure 8.5** The eight possible lowest energy structures assumed by the five most accessible fullerenes. Structures of (a) $C_{60}$-$I_h$, (b) $C_{70}$-$D_{5h}$, (c) $C_{76}$-$D_2$, (d) $C_{78}$-$D_3$, (e) $C_{78}$-$C_{2v}$(I), (f) $C_{78}$-$C_{2v}$(II), (g) $C_{84}$-$D_2$(IV), and (h) $C_{84}$-$D_{2d}$ (II). Reproduced with kind permission from Taylor and Burley, in *Fullerenes, Principles and Applications*, eds. F. Langa and J. F. Nierengarten, The Royal Society of Chemistry, Cambridge, 2007. Copyright Royal Society of Chemistry, 2007.

The possibility of isomerization in higher fullerenes while still satisfying the isolated pentagon rule can be understood along the notion of the formation of *Stone–Wales* rearrangement in the fullerene cage structure. The Stone–Wales rearrangement involves a 90° rotation of one of the carbon bonds in the structure. Figure 8.6 shows the interconversion of the $C_{78}$ molecule between the $C_2$ and $D_{3h}$ isomers via a Stone–Wales rearrangement.

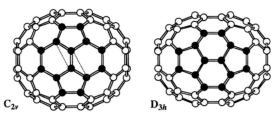

$C_{2v}$ $D_{3h}$

**Figure 8.6** Interconversion of the $C_{78}$ molecule between the $C_2$ and $D_{3h}$ isomers via a *Stone–Wales* rearrangement of a single carbon bond. Reproduced with kind permission from Diederich and Whetten, *Acc. Chem. Res.*, 1992, 25, 119. Copyright American Chemical Society, 1992.

The stability of some isomers of higher order fullerene is, indeed, an issue. For example, the $C_{2v}$ (II) isomer of [78] fullerene (see Figure 8.5) was reported to be stable for only 5 months after which it is completely degraded. No issues regarding the stability of $C_{60}$, or many of the other produced fullerenes, have been reported so far in inert environments. While the fact that fullerenes are typically produced in electric arcs with temperatures well above 4000 °C in inert atmospheres supports the notion that once produced, fullerenes are inherently stable. Some studies, however, showed that [60] fullerene is stable in low pressure of inert atmospheres only up to 950 °C. Issues have also been raised regarding the stability of $C_{60}$ in the presence of oxygen. While some recent studies based on infrared spectroscopy showed that the molecule is stable up to 600~650 °C, earlier studies based upon high-performance liquid chromatography suggested the degradation of $C_{60}$ into $C_{120}O$ even at room temperature in the presence of oxygen.

Out of the most accessible aforementioned fullerenes, the $C_{60}$ and $C_{70}$ are the most investigated members of the fullerene family. This most probably is due to their availability since the two isomers of fullerene are usually obtained in 75% and 24%, respectively, through the most common production method—the arc-discharge method of Krätschmer and Huffman. The reminder of the process product (1%) is usually a mixture of other forms of carbon including higher fullerenes reaching to higher that $C_{100}$. In the following section we will discuss the structures of these two most investigated fullerene molecules.

## 8.2.1   The Structure of the [60] Fullerene Molecule

Once described as the most beautiful molecule, $C_{60}$ or [60] fullerene has a unique structure. The fullerene cage for $C_{60}$ molecules has 12 pentagons and 20 hexagons. Figure 8.7 shows the [60] fullerene molecule. Inspection of the structure shows that each and every pentagon is surrounded by five hexagons. This makes the $C_{60}$ the smallest fullerene having no abutting pentagons.

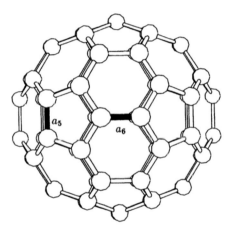

**Figure 8.7**   The structure of [60] fullerene molecules showing the two different types of C–C bonds; the "$a_5$" bordering a pentagon and a hexagon, and "$a_6$" bordering two hexagons.

Two types of C–C bonds have been distinguished[1] in the [60] fullerene structure (see Figure 8.7). These are the C–C bond bordering a hexagon and a pentagon, known in the literature as "$a_5$", and the C–C bond bordering two hexagons, known in the literature as "$a_6$." The lengths of these bonds were measured experimentally

---

[1]In some texts, these bonds are further distinguished and classified as single and double bonds. The author prefers to avoid such distinction due to the well-known fact that such sharp distinction is hard to make in aromatic structures.

by NMR and neutron diffraction methods. They were reported to range between 1.455 Å and 1.46 Å for the $a_5$ bond, and between 1.4 Å and 1.391 Å for the $a_6$ bond. The neutron diffraction method always yielded the lower value of the length range. Recently, it was shown that such bond length distribution matches the molecular simulation results for unperturbed $C_{60}$ molecule. Any perturbation of the molecule by an external thermodynamic field, however, was shown to dramatically alter the C–C bond length distribution in the fullerene molecule. Figure 8.8 shows the C–C bond distribution in a $C_{60}$ fullerene molecule interacting with different number of water molecules as determined by semi-empirical molecular simulation method. It is clear that the chemical potential of the interacting water molecules tremendously alters the bond length distribution within the fullerene cage. The results show that as the number of interacting molecules increases, the C–C bond length distribution within the fullerene molecule broadens and shifts to lower values. This would definitely have an impact on the fullerene properties as we will discuss later.

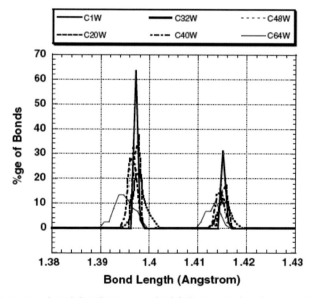

**Figure 8.8** C–C bond distribution in a [60] fullerene molecule interacting with different numbers of water molecules. Reproduced with kind permission from Amer et al., *Chem. Phys. Lett.*, 2005, 411, 395. Copyright Elsevier, 2005.

The C–C bonds in a [60] fullerene molecule (the vertices of the molecule) are usually considered in the literature to form a *regular* truncated icosahedron. In fact, this is an accepted approximation since the length difference between the $a_5$ and the $a_6$ bonds caused the molecule to slightly deviate from being a regular truncated icosahedron. The diameter of the [60] fullerene molecule is an important parameter of its structure to consider. Generally speaking, the diameter of an icosahedral fullerene ($d_i$) can be calculated based upon the number of carbon atoms ($n$) and the average C–C bond length ($a_{c-c}$) according to the equation;

$$d_i = \frac{5\sqrt{3}a_{c-c}}{\pi}\sqrt{\frac{n}{20}} \tag{8.4}$$

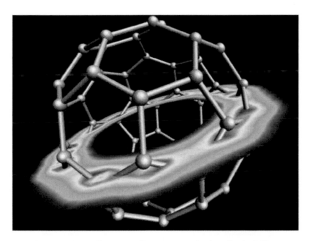

**Figure 8.9** The 1 nm effective diameter of the [60] fullerene molecule including the thickness of the $\pi$-electron shell associated with sp$^2$ hybridization of the carbon atoms. Reproduced with kind permission from Tycko et al., *J. Phys. Chem.*, 1991, 95, 518. Copyright American Chemical Society, 1991.

Using an approximated average value of 1.44 Å yields the diameter of the $C_{60}$ icosahedron to be 6.88 Å. However, it is important to note that the effective diameter of the $C_{60}$ molecule has to include the thickness of the $\pi$-electron shell associated with the sp$^2$ hybridization status of the carbon atoms forming the molecule (see Figure 8.9). Estimating the $\pi$-electron shell thickness by 3.37 Å, based on the very well characterized interplanar distance in graphite, put the effective diameter of a $C_{60}$ molecule very close to 1 nm. Recalling that the correlation length ($\xi$) is typically on the order of 1 nm, makes the fullerene molecule a perfect example for a nanosystem where the system size is on the same order of one of the system's characteristic lengths scales discussed earlier. This also means that all atoms making a $C_{60}$ molecule are correlated.

## 8.2.2 The Structure of the [70] Fullerene Molecule

The structure of next most investigated fullerene, the [70] fullerene, is slightly different from that of [60] fullerene. According to the Euler construction rule for polyhedra, five extra hexagons do exist in the structure of the [70] fullerene molecule compared to that of the $C_{60}$ molecule. Figure 8.10 shows the structure of the [70] fullerene molecule with the extra five hexagons clearly observed. These five extra hexagons extend the cage structure in one direction rendering a shape resembling that of a rugby ball. As we mentioned above, the $C_{70}$ molecule come in only one structure that belongs to the $D_{5h}$ symmetry point group.

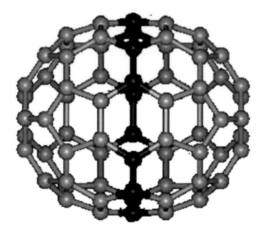

**Figure 8.10** The structure of [70] fullerene showing the extra 10 carbon atoms resulting in 5 extra hexagons.

Unlike the $C_{60}$ molecule, where all the carbon atom sites are equivalent, there are five different carbon atom sites in the $C_{70}$ molecule. This leads to 8 different types of C–C bonds in the molecular cage. Figure 8.11 shows the structure of the $C_{70}$ molecule elucidating the five nonequivalent carbon sites and the eight different C–C bonds connecting them. Theoretical calculations as well as experimental investigations were conducted to determine the C–C bond lengths in $C_{70}$. Table 8.3 shows examples of modeling and experimental results for the eight different C–C bonds in [70] fullerene as reported in the literature.

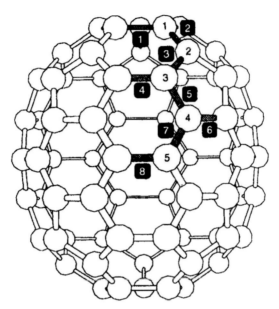

**Figure 8.11** The structure of [70] fullerene molecule elucidating the five nonequivalent carbon sites and the eight different C—C bonds. Reproduced with kind permission from Saito and Oshiyama, *Phys. Rev. B*, 1991, 44, 11532. Copyright American Physical Society, 1991.

As we will discuss later the 5 extra hexagons in the [70] fullerene molecule cause measurable differences in its properties and behavior compared to the properties and behavior of the [60] fullerene molecule. This is also true for the properties and behavior of other fullerenes as well. While the chemical composition of different fullerenes is essentially the same, the size, shape, and chirality of different types of molecules manifest themselves strongly in the properties and behavior of these molecules.

## 8.3 Endo-fullerenes

The internal cavity of [60] fullerene molecule dragged lots of attention. Giving our previous discussion, the internal cavity for $C_{60}$ would be on the order of 4 Å diameter cavity surrounded by a 3 Å–thick shell of electrons (see Figure 8.9). This triggered a line of investigation involving fullerenes (especially $C_{60}$) with incarcerated atoms. This class of fullerene is referred to as incar-fullerenes, endo-fullerenes, or endohedrals.

Different endo-fullerenes have been investigated theoretically and prepared experimentally. Incarceration of different elements in several types of fullerenes was investigated. The different elements investigated included nitrogen, hydrogen, noble gases (helium, neon, argon, krypton, and xenon), different metal nitrides, and different types of metallic, non-metallic, and even radioactive, atoms. Table 8.2 shows different endo-fullerenes reported in the literature.

**Table 8.2**  Examples of modeling and experimental results for the eight different C–C bonds in [70] fullerene as reported in the literature (see Figure 8.11 for bond locations)

| | Model | | Experiments | |
| | TBMD | LDA | Electron Diffraction | X-ray |
| Bond | | | | |
|---|---|---|---|---|
| $C_1$-$C_1$ | 1.457 | 1.448 | 1.46 | 1.434 |
| $C_1$-$C_2$ | 1.397 | 1.393 | 1.37 | 1.377 |
| $C_2$-$C_3$ | 1.454 | 1.444 | 1.47 | 1.443 |
| $C_3$-$C_3$ | 1.389 | 1.386 | 1.37 | 1.369 |
| $C_3$-$C_4$ | 1.456 | 1.442 | 1.46 | 1.442 |
| $C_4$-$C_4$ | 1.443 | 1.434 | 1.47 | 1.396 |
| $C_4$-$C_5$ | 1.418 | 1.415 | 1.39 | 1.418 |
| $C_5$-$C_5$ | 1.452 | 1.467 | 1.41 | 1.457 |

TBMD is tight binding molecular dynamic method, and LDA is local density analysis.

**Table 8.3**   Various endo-fullerenes investigated and reported in the literature

| Fullerene | Metal | References |
|---|---|---|
| | **Fullerenes incarcerating one atom** | |
| $C_{28}$ | Hf, Ti, U, Zr | 109 |
| $C_{36}$ | U | 109 |
| $C_{44}$ | K, La, U | 109,112 |
| $C_{48}$ | $C_S$ | 112 |
| $C_{50}$ | U | 109 |
| $C_{60}$ | Li, K, Ca, Co, Y, Cs, Ba, Rb, La, Ce, Pr, Nd, Sm, Eu, Gd, Tb, Dy, Ho, Er, Lu, U | 110,114 |
| $C_{70}$ | Li, Ca, Y, Ba, La, Ce, Gd, Lu, U | 109,111,114 |
| $C_{72}$ | U | 109 |
| $C_{74}$ | Sc, La, Gd, Lu | 112,113,115 |
| $C_{76}$ | La | 109,115 |
| $C_{80}$ | Ca, Sr, Ba | 116 |
| $C_{82}$ | Ca, Sc, Sr, Ba, Y, La, Ce, Pr, Nd,Sm, Eu, Tm, Lu | 109,113,115 |
| $C_{84}$ | Ca, Sc, Sr, Ba, La | 115,119,120 |
| | **Fullerenes incarcerating two atoms** | |
| $C_{28}$ | $U_2$ | 109 |
| $C_{56}$ | $U_2$ | 109 |
| $C_{60}$ | $Y_2$, $La_2$, $U_2$ | 109 |
| $C_{74}$ | $Sc_2$ | 118 |
| $C_{76}$ | $La_2$ | 115 |
| $C_{80}$ | $La_2$, $Ce_2$, $Pr_2$ | 117,110 |
| $C_{82}$ | $Sc_2$, $Y_2$, $La_2$ | 117,115 |
| $C_{84}$ | $Sc_2$, $La_2$ | 113,115 |
| | **Fullerenes incarcerating three atoms** | |
| $C_{82}$ | $Sc_3$ | 115 |
| $C_{84}$ | $Sc_3$ | 121 |
| | **Fullerenes incarcerating four atoms** | |
| $C_{82}$ | $Sc_4$ | 116 |

*(Continued)*

**Table 8.3** (*Continued*)

| Fullerenes incarcerating derivatives of ammonia $NR_3$, $NR_2R'$, or $NRR'R''$ | | |
|---|---|---|
| $C_{78}$ | $Sc_3N$ | 123 |
| $C_{76}$ | $Dy_3N$ | 122 |
| $C_{68}$ | $Sc_3N$, $Sc_2(Tm/Er/Gd/Ho/La)_2N$ | 124 |
| $C_{80}$ | $Sc_3N$, $ErSc_2N$, $Er_2ScN$, $Lu_3N$, $Lu(Gd/Ho)_2$, $Y_3N$, $Ho_3N$, $Tb_3N$, $Dy_3N$ | 122,125,126 |
| $C_{82\text{-}98}$ | $Dy_3N$ | 122 |

Endo-fullerenes, in general, and the easy to produce *incartrimetalnitridofullerenes*, in particular, have opened a new gate to other unchartered territories. The technological and biomedical applications of such class of fullerenes is very promising and still to be explored. Figure 8.12 shows different types of endo-fullerene molecules in which an atom, multiple atoms, or a molecule are incarcerated in a fullerene cage of variable sizes or shapes.

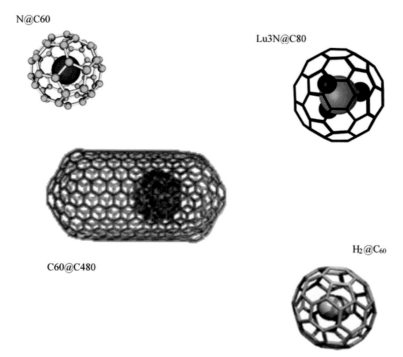

N@C60

Lu3N@C80

C60@C480

$H_2@C_{60}$

**Figure 8.12**   Different types of endo-fullerene molecules.

## 8.4    Fullerene Onions

Fullerene nano-onions are quasi-spherical particles consisting of concentric graphitic shells. The nano-onions were first observed when carbon soot particles and tubular graphitic structures were exposed to intense electron beam irradiation in transmission electron microscopy experiment. Apparently, the extremely high local temperature within the electron beam allowed structural fluidity leading to the formation of the new form of carbon nanospecies. Nano-onions were also observed as the result of nanodiamond particles annealing in vacuum at temperatures ranging between 1000 °C and 1500 °C, and also upon generating an arc discharge between carbon electrodes submerged in water. Figure 8.13 shows a

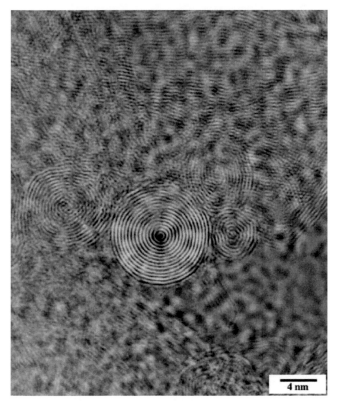

**Figure 8.13**    High resolution transmission electron micrograph showing carbon nano-onions produced by arc discharge between carbon electrodes submerged in water. Reproduced with kind permission from Roy et al., *Chem. Phys. Lett.*, 2003, 373, 52. Copyright Elsevier, 2003.

high resolution transmission electron micrograph of a carbon nano-onion. As clear from the figure, the carbon nano-onions consist from several concentric shells. Up to 15 shells can be counted in a nano-onion with roughly a 12 nm diameter produced by the submerged carbon arc method.

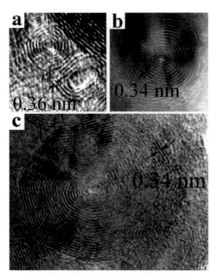

**Figure 8.14** Carbon onions formed in graphite single crystals under pressure and shear deformation: (a) an onion with 6–7 layers (24 GPa pressure before shear and 46 GPa after shear, 15 shear cycles), (b) an onion with 14–15 layers (48 GPa pressure before shear and 63 GPa after shear, 20 shear cycles), (c) an onion with about 30 layers (57 GPa pressure before shear and 71 GPa after shear, 5 shear cycles). Reproduced with kind permission from Blank et al., *Nanotechnology*, 2007, 18. Copyright Institute of Physics, 2007.

More recently, carbon onions as large as 2 μm in diameter were observed after high sheer and hydrostatic loading of graphite single crystals in a diamond anvil cell (DAC) at room temperature. The number of layers, and hence, the size of the carbon onion was found to increase with increasing the loading pressure. Small onions with 4–6 layers formed at lower pressure ranges (46 GPa after shear) and large onions with up to 60 layers formed under higher pressures (71 GPa after shear). Figure 8.14 shows high resolution transmission electron micrographs of carbon onions observed after high sheer and hydrostatic pressure treatment at room temperature. Interestingly enough, the shell to shell distance was found to range between 0.317

nm and 0.36 nm for different size onions with the smaller distance characteristic of the larger onions. Figure 8.15 shows a scanning electron micrograph of the giant carbon onions observed as a result of this treatment of single graphite crystals.

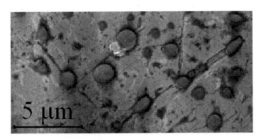

**Figure 8.15**   Scanning electron micrograph of giant carbon onions about 2 μm in diameter. Reproduced with kind permission from Blank et al., *Nanotechnology*, 2007, 18. Copyright Institute of Physics, 2007.

Carbon onions have also been produced and observed by many other means such as carbon soot annealing, regular arc discharge, carbon ion beam implantation on silver and copper substrates, plasma spraying of nanodiamond, catalytic decomposition of methane over an Ni/Al catalyst, as well as after $C_{60}$ thermobaric treatment. In addition, Carbon onions having an amorphous-like silicon carbide (SiC) core have been produced from polycrystalline SiC powder subjected to laser shock compression. Carbon onions have been also shown to exist in the interstellar dust.

Our state of knowledge regarding carbon nano-onions can be safely described as being on at the start of a steep learning curve. In spite of the fact that a number of studies have been done to investigate the production, formation mechanism, and properties of carbon nano-onions, the field is virtually unexplored and there is still plenty of space to fill regarding the properties and applications of such a unique class of carbon nanobuilding blocks.

## 8.5   Giant Spherical Fullerenes

A close look at the laser TOF mass spectroscopy (LTOF-MS) results presented in Figure 8.22 reveals that experimental extraction, and hence observation, of fullerenes with more than ~110 carbon atoms are not common. As a matter of fact, the largest experimentally extracted fullerene reported contained about 212–266 carbon atoms according to LTOF-MS results. In an attempt to answer the question "are giant fullerenes spherical or tubular?" theoretical investigation using ab initio calculations were conducted to calculate the energy of higher fullerenes (ranging from $C_{80}$ up to $C_{240}$) in both spherical (icosahedral) and tubular (cylindrical) geometries as shown in Figure 8.16. The results showed that in all investigated fullerenes the icosahedral geometries are much more stable than the cylindrical "buckytube" geometries. In addition, scanning tunneling microscopy images of fullerenes with ~300 carbon atoms according to LTOF-MS results showed that these giant fullerenes have roughly spherical shapes with diameters ranging between 1 nm and 2 nm with no evidence of tube-like geometries.

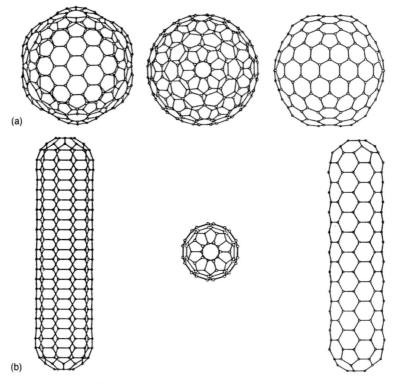

(a)

(b)

**Figure 8.16** Equilibrium geometries of $C_{240}$ as determined by ab initio calculations; (a) spherical $I_h$ geometries, and (b) $D_{5d}$ cylindrical geometries. Reproduced with kind permission from G. Scuseria, *Chem. Phys. Lett.*, 1995, 243, 193. Copyright Elsevier, 1995.

These results and the fact that theoretical investigations predicted the possible stability of fullerenes with up to 4860 carbon atoms as we mentioned before, raises a very interesting question regarding the ability to extract and observe such gigantic fullerenes. The answer for such question has been proposed along the notion that such giant fullerenes could be either forming in amounts too small to observe experimentally or having a solubility, in the typically used extraction solvents, too low to enable their extraction. Another plausible reason for the discrepancy is the fact that all simulations are done in absolute vacuum environment. Processing, however, is not and the effect of such environment could be the reason that giant fullerene while shown more stable than tubular in the calculations are not observed in experimental. The other two possibilities for geometries of fullerenes with giant masses are either tubular or in the form of fullerene onions as we will discuss in the following chapters.

## 8.6    Production Methods of Spherical Fullerenes

Fullerenes can be generally produced in laboratory facilities in different ways involving the generation of a carbon-rich vapor or plasma. There are essentially three methods—with many modifications—to produce 0-D fullerenes. These methods are; the Huffman–Krätschmer, or the arc-discharge process, benzene combustion in oxygen deficient environment, and condensation of polycyclic aromatic hydrocarbons.

### 8.6.1    The Huffman–Krätschmer Method

As we mentioned earlier, this method was the earliest method to produce fullerenes in significant amounts and its development, actually, marked the beginning of the fullerene science. The Huffman–Krätschmer method involves arc discharge between highly pure carbon rods in a helium or argon atmosphere at 100~200 torr of pressure. A mechanical mechanism is needed to translate the electrodes together in order to maintain the electrode gap as they get consumer. Controlling such a gap was reported to be essential for the process and to prevent temperature drop. At estimated electrode tip temperature of 2000 °C, the yield of fullerene in the produced soot is about 4%. However, at estimated electrode tip temperature of ~4700 °C, the yield of fullerene in the produced soot was reported to increase into the 7–10% range. It is important to note that yield calculations tend to account for all types of formed fullerenes. $C_{60}$ and $C_{70}$, however, constitute the majority in the produced fullerene.

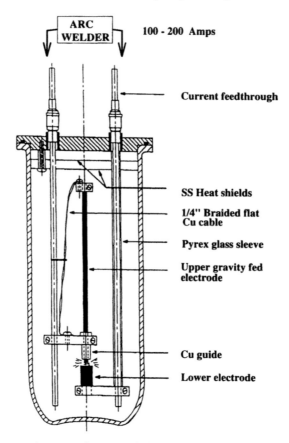

**Figure 8.17** Schematic diagram of the bench-top contact arc apparatus. Reproduced with kind permission from Koch et al., *J. Org. Chem.*, 1991, 56, 4543. Copyright American Chemical Society, 1991.

A simple bench-top reactor was developed to produce fullerene with 4% yield as well. The reactor utilizes an inexpensive ac arc welding power supply to initiate and maintain a contact arc between two graphite electrodes in a helium atmosphere. Since this technique does not require a mechanism to translate the electrodes together as they get consumed, it is termed the "contact arc method." Instead, the technique utilizes flexible support for the upper electrode and relies on gravity to maintain contact between the two vertical electrodes. Figure 8.17 shows a schematic diagram of the bench-top contact arc apparatus.

In rather a big leap forward, Parker et al. used a plasma arc in a fixed gap between two horizontal electrodes. The developed apparatus is known as *fullerene generator* due to the high yield it generates reaching ~40% in high boiling point solvents. The exceptionally high yield of this method was attributed to the fine control of the arc gap combined with proper convection of the atmosphere in the apparatus and carful extraction. Figure 8.18 shows a schematic diagram of the plasma arc *fullerene generator*.

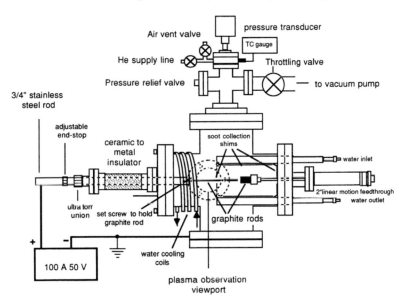

**Figure 8.18** A schematic diagram of the plasma arc fullerene generator. Reproduced with kind permission from Parker et al. in *The Fullerenes*, eds., H. Kroto, J. E. Fisher, and D. E. Cox, Pergamon Press, Oxford, 1993, p. 29. Copyright Pergamon Press, 1993.

Several processing parameters are currently known to affect the fullerene yield in an arc-discharge process. The optimum atmosphere to be used was found to be highly pure helium. While most reactors are operated in the 100–200 torr level, the optimum operation pressure was reported to be highly sensitive to the actual chamber design and should be determined for each specific reactor. Although the purity of the carbon electrodes was found not to affect the soot production rate, the smaller-diameter electrodes were found to have higher yield of soot. In addition, it was found that small contaminations of hydrogen or moisture in the generation chamber would seriously suppress fullerene generation.

### 8.6.2   The Benzene Combustion Method

Evidence that fullerenes can be formed in flames was at first elusive, but progress was eventually made. In 1991 significant quantities of $C_{60}$ and $C_{70}$ were found in samples collected from low-pressure premixed benzene/oxygen flames. Further investigations showed that fullerenes can be produced in substantial quantities by sub-atmospheric pressure, laminar, premixed flames of benzene in oxygen deficient atmosphere with or without the presence of an inert gas. The largest yield of soot into fullerene was reported to be 20% at a pressure of 37.5 torr. Figure 8.19 shows a schematic diagram of the burner and associated equipment used in the process. It was also reported that fullerene formation in the flame can take place with the presence of hydrogen and oxygen. The promise of the combustion method encouraged Frontier Carbon Corporation, a subsidiary of Mitsubishi Chemical Corporation, to construct a large scale fullerene factory in Japan in 2003. The factory has the capacity of producing 5000 ton of fullerene annually.

### 8.6.3   The Condensation Method

The condensation method is based upon condensation of polycyclic aromatic hydrocarbons through pyrolytic dehydrogenation or dehydrohalogenation processes. While the method does not produce fullerenes in sufficient quantities for practical applications, it provides a means to deduce the mechanism of fullerene formation. The method was used to produce only $C_{60}$ fullerene from a molecular polycyclic aromatic precursor bearing chlorine substituents at key positions subjected to flash vacuum pyrolysis at 1100 °C through a 12 steps reaction. Figure 8.20 shows a schematic for the precursor and the formed $C_{60}$.

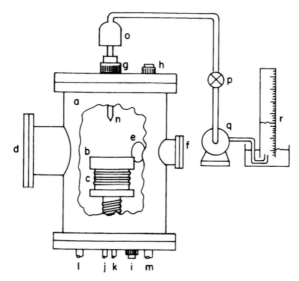

**Figure 8.19** A schematic diagram of the burner and associated equipments used in the combustion process, a— low pressure chamber; b— copper-burner plate; c— water cooling coil; d, e, and f— windows; g, h, and i— feedthroughs; j— annular flame feed tube; k— core-flame feed tube; l and m— exhaust tubes; n— sampling probe; o— filter; p— valve; q— vacuum pump; r— gas meter. Reproduced with kind permission from Howard et al., *Nature*, 1991, 352, 139. Copyright Nature Publishing Group, 1991.

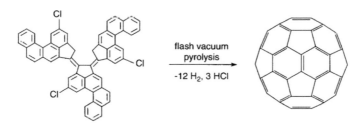

**Figure 8.20** Formation of [60] fullerene from chloroaromatic precursor through the condensation process. Reproduced with kind permission from Scott et al., *Science*, 2002, 295, 1500. Copyright AAAS, 2002.

## 8.7 Extraction Methods of Fullerenes

The main method used to extract fullerene from the produced soot is the traditional Soxhlet extraction method. The Soxhlet method is traditionally used to extract molecular moieties from solid phases using organic solvents capable of dissolving the molecular moieties. Figure 8.21 schematically illustrates a Soxhlet extraction unit. The solid sample containing the molecular moiety to be extracted is loaded into a thimble in the Soxhlet extractor. As the solvent is boiled in the flask at the bottom, solvent vapors rise through the side channel on the left of the extractor and condense near the bottom of the condenser unit, resulting in dripping hot distilled solvent into the thimble through the solid sample. The hot distilled solvent extracts the molecular moiety on its way down through the solid sample, and the solution, then, makes its way back to the flask via the tube to the right. This closed-loop system is usually operated for several hours during which all extractable molecular moieties are collected in the flask. The non-extractable portion of the solid sample remains in the thimble.

Fullerenes are extracted from the produced soot using the Soxhlet apparatus with any of different types of solvents. Many solvents have been used in the extraction process such as chloroform, toluene, benzene, $n$-hexane, 1,2-dichlorobenzene, etc. The type of the solvent controls the speed of the extraction process and dictates subsequent processes. For example, chloroform results in a very slow process. The use of 1,2-dichlorobenzene results in a very fast extraction process but requires high vacuum process for removal of solvent traces. In addition, if carbon disulfide is used, it has to be vigorously removed under vacuum to avoid fullerene contamination with sulfur. Selective extraction of various molecular weight fullerenes by varying the extraction solvent was also reported. Higher mass fullerenes are better extracted with more polar and higher boiling point solvents. Figure 8.22 shows a fullerene extraction and separation scheme. The laser desorption time of flight (TOF) mass spectra of soot extracts are also shown in the figure. The results shown in the figure are for soot produced using the combustion method described earlier.

The figure highlights two important points. First, the total fullerene yield is sharply dependent of the extraction scheme and solvents used. Secondly, the extracted fullerene molecular weight depends on the type of solvent used in the extraction process. For example, while benzene mainly extracts $C_{60}$ and $C_{70}$ with smaller amounts of higher fullerenes up to $C_{96}$, 1,2,3,5-tetramethylbenzene (TMB) extracts more of the higher fullerenes. In addition, it is clear from the figure that while hexane mainly extracts $C_{60}$ and $C_{70}$, heptane additionally extracts $C_{80}$ and $C_{84}$.

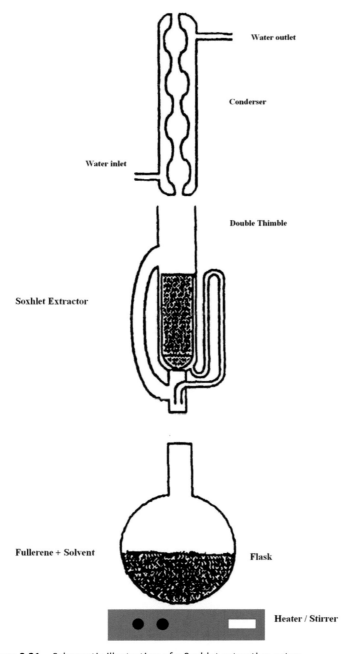

**Figure 8.21**  Schematic illustration of a Soxhlet extraction setup.

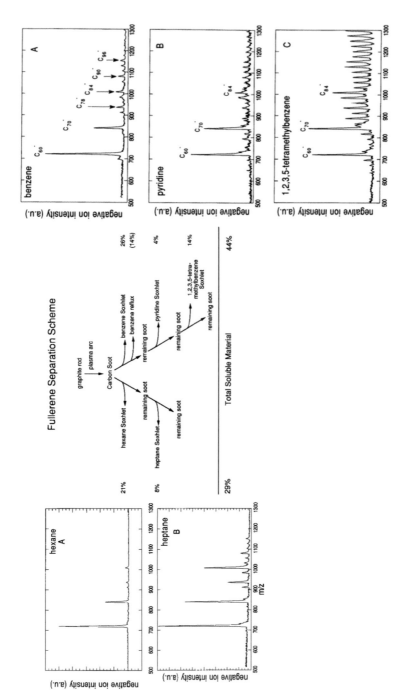

**Figure 8.22** Fullerene extraction and separation scheme along with Laser TOF mass spectra of extract prepared by sequential solvent extraction. Adapted with kind permission from Parker et al. in *The Fullerenes*, eds., H. Kroto, J. E. Fisher, and D. E. Cox, Pergamon Press, Oxford, 1993, p. 29. Copyright Pergamon Press, 1993.

An efficient alternative method has also been proposed in the literature where the produced carbon soot is dispersed in tetrahydrofuran (THF) at a concentration of 0.1 g/ml at room temperature and then sonicated for 20 minutes. After filtration to remove insoluble remains, the THF is removed using a rotary evaporator operated at 50 °C, leaving fullerenes and other soluble impurities in the flask. In this case, the impurities contain many polyaromatic hydrocarbons that can be removed by washing the extract in diethyl ether before it is further purified.

A non-solvent based method, the sublimation method, was also utilized for fullerene extraction from the produced soot. $C_{60}$ and $C_{70}$ powders are known to sublime in vacuum at relatively low temperatures (i.e., 350 °C and 460 °C, respectively). The advantage of the sublimation extraction method is that it produced fullerene samples that are solvent free. This is crucial for experiments were any traces of solvent could greatly affect the investigation results. In the sublimation method, the soot sample is placed in one end of an evacuated quartz tube placed in a furnace with a temperature gradient. The tube end containing the soot sample is kept at the highest temperature zone of the furnace (600–700 °C). Fullerenes will sublime and drift down the temperature gradient to condense on the walls of the quartz tube. The higher the fullerene mass, the closer to the soot it will condense.

## 8.8 Purification Methods of Fullerenes

Purification of fullerenes is intended to separate fullerene molecules from any impurities such as polyaromatic hydrocarbons and other types of carbon-based impurities. In addition, purification process is used to separate fullerenes from each other based on their molecular mass, and size. Mainly solvent methods based on liquid chromatography (LC) and sublimation methods based on sublimation under temperature gradients are used for fullerene purification. Sublimation method for purification is essentially the same at the sublimation method for extraction we have discussed earlier. The effectiveness of the purification process is usually verified by fullerene sensitive characterizations techniques such as mass spectroscopy, nuclear magnetic resonance, and infrared, Raman, or optical absorption spectroscopy.

*Liquid chromatography* is the main technique for fullerene purification. In this technique, a solution containing a mixture of fullerenes to be separated (purified) (usually referred to as the mobile phase) is forced through a column filled with a high-surface area solid (usually referred to as the stationary phase). As the mixed solution passes through the column, different molecules in the solution experience different levels of interaction with the high-surface area solid in the column due to various physical and chemical mechanisms. Stronger interactions increase the retention time of the molecule in the column, or in other words, decrease the migration rate of that particular molecule through the column. Hence, separated molecules should elude from the column in the order of decreasing retention time. Weakly interacting molecules will have shorter retention times and will elude first. Strongly interacting molecules will have longer retention times and will elude later.

In case of fullerenes, the identity of the separated fullerene is verified qualitatively by color and more quantitatively by other characterization techniques mentioned above. Regarding the quantitative identification, different fullerenes are identified by the color of their solution in certain solvents. For example, in toluene, $C_{60}$ yields a magenta or purple solution, while $C_{70}$ yields a reddish to orange solution. Liquid chromatography generally enables the separation of fullerenes based upon their size (or mass).

Effective separation (or purification) can be achieved when a sufficient difference in the retention time can be achieved. This depends on the nature of the "stationary phase" used in the process. Different types of stationary phases have been used. The most widely used were based on silica gels and alumina. Carbon based stationary phases (Elorite carbon) were proved efficient in firmly holding higher fullerene while allowing [60] and [70] fullerenes to elute.

Fullerene purification procedures combining the extraction and purification steps were also developed. The procedure is based on a modified Soxhlet technique discussed earlier. Figure 8.23 shows a schematic illustration of the modified Soxhlet apparatus which is reported to be capable of extracting 1 g of $C_{60}$ and 0.1 g of $C_{70}$ a day. As shown in the figure, the thimble in a regular Soxhlet apparatus has been replaced with a liquid chromatography column which is filled with neutral alumina. During operation, the distilled solvent (pure hexane) is condensed at the top of the column to pass down the column extracting $C_{60}$ molecules into the distillation flask. After about 20–30 hours of operation, the flask containing $C_{60}$ in hexane is replaced with a second pure hexane flask and $C_{70}$, remaining in the column, is collected in the same way.

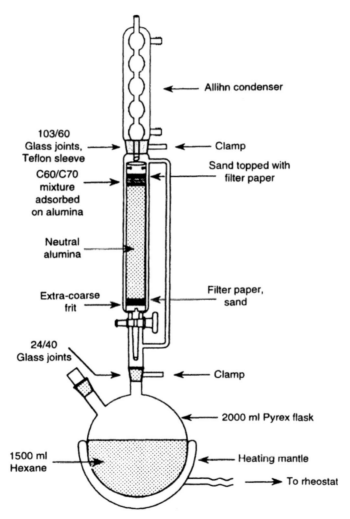

**Figure 8.23** A schematic illustration of the modified Soxhlet apparatus which combines the separation and purification steps. Reproduced with kind permission from Khemani et al., *J. Org. Chem.*, 1992, 57, 3254. Copyright American Chemical Society, 1992.

## 8.9 Properties of 0-D Fullerenes

In this section, the unique properties of the spherical fullerene molecules will be discussed. Such properties will include their spectroscopic properties (Raman spectroscopy), mechanical properties, electric properties. Their interaction with solvents will be discussed as well including the effect of these fullerenes on the solvent and the effect of solvents on these fullerenes.

### 8.9.1 The Raman Scattering of Fullerenes

Having a dimensionality on the order of 1 nm, spherical fullerenes demonstrated unique light scattering, in general, and Raman scattering, in particular, properties. In fact, Raman spectroscopy plays a crucial role in addressing fundamental questions regarding the physics of fullerenes. Critical issues such as the structure and properties of fullerenes and their molecular crystals have been addressed and largely understood based on Raman investigations. Raman spectroscopy has also been utilized to investigate thermodynamic equations-of-state and phase transitions in certain fullerene molecular solids. In addition, surface interactions and adsorption behavior of $C_{60}$, and other fullerene forms, was greatly elucidated based on Raman scattering studies.

#### 8.9.1.1 Raman Scattering of C60 Molecules and Crystals

As we mentioned before a [60] fullerene molecule ($C_{60}$) has 60 carbon atoms in a cage-like structure belonging to the icosahedral ($I_h$) point group symmetry. Due to the symmetry of the $C_{60}$ molecule, its 174 normal vibration modes ($3N-6$) can be reduced to 46 distinct symmetry species according to the following symmetries:

$$\Gamma = 2A_g + 3F_{1g} + 4F_{2g} + 6G_g + 8H_g + A_u + 4F_{1u} + 5F_{2u} + 6G_u + 7H_u. \quad (8.5)$$

Only ten of these vibrational modes, two belonging to the $A_g$ ($2A_g$) and eight belonging to the $H_g$ species ($8H_g$) are Raman active. The 4 ($F_{1u}$) modes are infrared active, while the remaining 32 modes are optically silent. Three of the Raman active modes [$H_g$ (7), $A_g$ (2), and $H_g$ (8)] are surface modes. They correspond to Pentagon shear, pentagon pinch, and hexagon shear modes, respectively. The $A_g$ (1) is a radial "breathing mode where all atoms are radially displaced at equal magnitude and in phase. Figure 8.24 schematically shows the atomic displacement vectors corresponding to the 10 Raman active modes of the $C_{60}$ molecules.

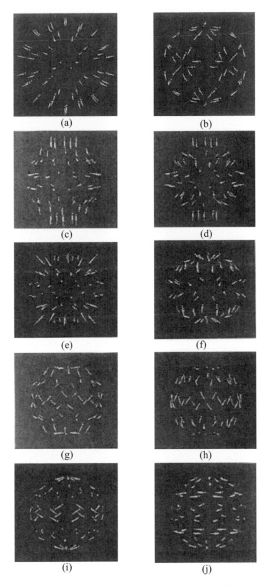

**Figure 8.24** Schematic diagram depicting the atomic displacement vectors corresponding to Raman active vibration modes in $C_{60}$. (a) $A_g$ (1), (b) $A_g$ (2), (c) $H_g$ (1), (d) $H_g$ (2), (e) $H_g$ (3), (f) $H_g$ (4), (g) $H_g$ (5), (h) $H_g$ (6), (i) $H_g$ (7), and (j) $H_g$ (8). Reproduced with kind permission from Schlüter et al., *J. Phys. Chem. Solids*, 1992, 53, 1473. Copyrights Elsevier, 1992.

Such vibration modes are known as the intra-molecular (or simply, molecular) vibration modes since they generate from atomic degrees of freedom within the molecule itself. Later, we will discuss other type of vibration modes (referred to as inter-molecular, or lattice, vibration modes) that has been observed in groups of $C_{60}$ fullerene molecules interacting and forming crystals or cooperative structures. The energies, and hence, the frequencies, of the Raman active intra-molecular vibration modes of $C_{60}$ have been theoretically investigated and calculated as early as 1987. Experimental measurements of $C_{60}$ Raman spectrum, however, had to wait until the Krätschmer–Huffman method was developed in 1990 due to sample availability issues. For the past twenty years, the Raman activity of $C_{60}$ has been extensively investigated both theoretically and experimentally. Table 8.4 shows the theoretical versus the experimental frequencies of the Raman active modes for $C_{60}$ molecule. Phonon calculations and Polarized Raman scattering studies of $C_{60}$ solid films showed that in the ($\parallel/\parallel$) polarization arrangement, both $A_g$ and $H_g$ modes are observable. In the ($\parallel/\perp$) polarization setup, however, only the $H_g$ modes can be experimentally observed. Figure 8.25 shows experimental Raman spectrum of $C_{60}$ obtained from a solid $C_{60}$ film on silicon substrate.

**Table 8.4**  Symmetry assignment, theoretical and experimental frequencies of Raman active modes in $C_{60}$ molecule

| | | Theoretical Calculations | |
| --- | --- | --- | --- |
| Mode | Frequency | Giannozzi and Baroni [45] | Adams et al. [31] |
| $H_g$ (1) | 272 | 259 (−4.8%) | 259 (−4.8%) |
| $H_g$ (2) | 433 | 425 (−1.8%) | 427 (−1.4%) |
| $A_g$ (1) | 496 | 495 (−0.2%) | 494 (−0.4%) |
| $H_g$ (3) | 709 | 711 (−0.3%) | 694 (−2.1%) |
| $H_g$ (4) | 772 | 783 (−1.4%) | 760 (−1.6%) |
| $H_g$ (5) | 1099 | 1020 (−1.9%) | 1103 (0.4%) |
| $H_g$ (6) | 1252 | 1281 (−2.8%) | 1328 (6.1%) |
| $H_g$ (7) | 28825 | 28852 (−1.8%) | 1535 (7.7%) |
| $A_g$(2) | 28870 | 1504 (−2.3%) | 1607 (9.3%) |
| $H_g$ (8) | 1575 | 1578 (0.2%) | 1628 (3.4%) |

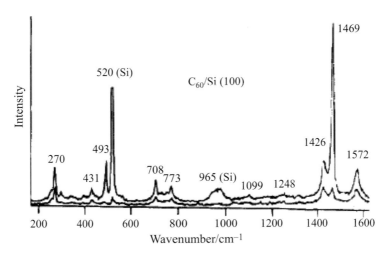

**Figure 8.25** Polarized Raman spectra of $C_{60}$ on a silicon substrate. Upper trace is for (∥/∥) polarization which shows both $A_g$ and $H_g$ modes. Lower trace is for (∥/⊥) polarization that only shows $H_g$ modes. Reproduced with kind permission from Eklund et al., *J. Phys. Chem. Solids*, 1992, 53, 1391. Copyright Elsevier, 1992.

It is clear from Table 8.4 that the experimentally measured frequencies are always different from their theoretically calculated values. While most of the theoretical calculations studies tend to judge the accuracy of their results by comparing them to experimental values, it should be emphasized that the unavoidable presence of perturbation fields around the $C_{60}$ molecule under experimental investigations renders such comparison meaningless. Calculated Raman frequencies are performed on individual isolated molecules. Meanwhile, experimental measurements are performed on fullerenes in a condensed state in the form of solid crystals (*fullerite*) or thin sublimated films. Perturbation fields resulting from the chemical potentials of neighboring molecules, molecules of the testing environment, and testing temperatures should all have a measurable effect on the Raman spectrum of the tested molecules.

In the solid state, $C_{60}$ molecules assume a simple cube (SC) unit cell which exhibit a phase transition into a face center cubic (FCC) upon heating around a transition temperature $T_{01} \approx 260$ K. The transition is also associated with restriction of rotation of the $C_{60}$ molecules. Above the transition temperature fullerene molecules are rotating almost freely in the FCC unit cell. Once the "orientational ordering temperature" is reached, rotation of the fullerene molecules is restricted to rotation about one of the {111} directions in the cell as shown schematically in Figure 8.26.

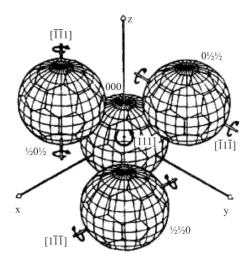

**Figure 8.26** The four molecules in the FCC unit cell of molecular $C_{60}$ solid. In this orientation, molecules at (0,0,0), (½,½,0), ((½,0,½), and (0,½,½) rotate by the same angle about the local axes [111], [1,$\bar{1}$,$\bar{1}$], [$\bar{1}$,$\bar{1}$,1], and [$\bar{1}$,1,$\bar{1}$], respectively. The sense of rotation about each axis is shown. Reproduced with kind permission from Copley et al., *Neutron News*, 1993, 4, 20. Copyright, Taylor and Francis Publishers, 1993.

In the solid state, $C_{60}$ is nearly an ideal molecular solid with weak intermolecular van der Waals interactions. This fullerene–fullerene interaction and the associated symmetry change (due to crystal formation) induce a new set of intermolecular of lattice vibration modes or phonons. Generally, lattice vibrational modes can be classified into acoustic, optical, or liberational modes. Optical 'lattice' modes are found in all solids with primitive unit cells containing more than two atoms or molecules. Hence, such optical modes are expected in the case of FCC $C_{60}$ crystals. Liberational modes, however, are specific for molecular crystals. They typically originate from the inertia of the individual molecule and are associated with a hindered rotation (rocking) of the molecules about their equilibrium lattice sites. For $C_{60}$ molecular crystals, the frequencies of the liberational and lattice modes are very low (less than 40 cm$^{-1}$ and less than 100 cm$^{-1}$, respectively) because of the weak molecular coupling (interaction) and the large molecular moment of inertia of fullerene molecules. While some experimental investigations attributed

9 new modes, with frequencies ranging between 300 cm$^{-1}$ and 1620 cm$^{-1}$, observed during their investigation of pressure induced phase transition in $C_{60}$ crystals to phonon modes, the frequencies are too high to fit the interpretation. Interaction between fullerene molecules and the solvent used as a pressure transmission medium in the study is most plausibly the reason for the observed new modes. Solvent interactions and solubility of the fullerenes are discussed in a following section. During the "fullerene rush" in the early 1990s of last century, several investigations led to conclusions that were later revised, or are still a controversy. For example, the origin of the non-polarized 1458 cm$^{-1}$ shoulder sometimes observed in the $C_{60}$ spectrum under different measurement conditions had several interpretations. Such interpretations included association with a downshift of the intrinsic pentagonal pinch mode, association with a photo-polymerized state of fullerene, association with oxygen adsorption and the presence of two different phases of the fullerite crystal. This controversy, and many others as we will discuss later, is still not completely resolved.

The power of Raman spectroscopy is best demonstrated in its ability to provide information regarding the presence of carbon isotopes ($^{13}$C) within the fullerene molecule. The isotope $^{13}$C exists with natural abundance of 1.1%. Hence for $C_{60}$ molecules, almost half of the molecules would have one or more $^{13}$C isotope atoms in their structure. Due to the presence of the isotope atom(s) within the structure, the symmetry of the fullerene will be lowered with expected effect on Raman modes intensity and position. We have mentioned earlier that the frequency of a vibration mode (v) is actually scaled to the reduced mass of the atoms involved in the mode. Hence, the presence of one or two heavier isotopes in the fullerene cage is expected to lower the frequency of the mode involving such isotopes. Experimentally, Raman lines associated with $^{12}C_{60}$, $^{13}C_{1}{}^{12}C_{59}$, and $^{13}C_{2}{}^{12}C_{58}$ could be resolved with a separation of $1 \pm 0.02$ cm$^{-1}$ between these Raman peaks. Figure 8.27a shows high resolution non-polarized Raman spectrum around the 1469 cm$^{-1}$ "pentagon pinch" mode of a frozen solution of $C_{60}$ in $CS_2$ at 30 K. The spectrum shows a three-Lorentzian fit to the experimental data. The highest wavenumber peak is assigned to the totally symmetric pentagonal pinch $A_g(2)$ mode in $^{12}C_{60}$. The other two lines are assigned to the pentagonal pinch mode in molecules containing one and two $^{13}$C atoms, respectively. Figure 8.27b shows the measured nonpolarized Raman spectrum in the pentagonal-pinch region for a frozen solution of $^{13}C$ enriched $C_{60}$ in CS2 at 30 K (points) as well as the theoretical spectrum computed using the sample's mass spectrum (solid line).

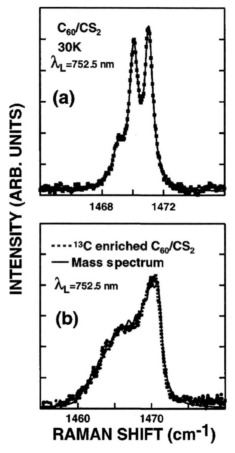

**Figure 8.27** (a) High resolution non-polarized Raman spectrum around the 1469 cm$^{-1}$ "pentagon pinch" mode of a frozen solution of $C_{60}$ in $CS_2$ at 30 K. The spectrum shows a three-Lorentzian fit to the experimental data. The highest wavenumber peak is assigned to the totally symmetric pentagonal pinch $A_g(2)$ mode in $^{12}C_{60}$. The other two lines are assigned to the pentagonal pinch mode in molecules containing one and two $^{13}C$ atoms, respectively. (b) Measured unpolarized Raman spectrum in the pentagonal-pinch region for a frozen solution of $^{13}C$ enriched $C_{60}$ in $CS_2$ at 30 K (points) as well as the theoretical spectrum computed using the sample's mass spectrum (solid line). Reproduced with kind permission from Guha et al., *Phys. Rev. B*, 1997, 56, 15431. Copyright American Physical Society, 1997.

Higher order Raman bands including overtones and recombination in $C_{60}$ were also experimentally observed. Figure 8.28 shows the spectrum recorded in the spectral range 100–3500 cm$^{-1}$.

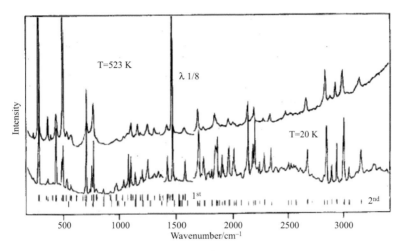

**Figure 8.28** Full range Raman spectra of $C_{60}$ showing all first order in addition to overtones and combinations modes of $C_{60}$ at low (T = 20 K) and high (T = 523 K) temperatures. Reproduced with kind permission from Dong et al., *Phys. Rev. B*, 1993, 48, 2862. Copyright American Physical Society, 1993.

### 8.9.1.2 Raman scattering of $C_{70}$

Raman scattering of $C_{70}$ was also investigated extensively both theoretically and experimentally. The Raman spectrum of $C_{70}$ is much more complicated than that of $C_{60}$ because of the lower symmetry of the $C_{70}$ molecule ($D_{5h}$), and contains 53 Raman active modes. Figure 8.29 shows one of the earliest reported nonpolarized Raman spectrum of $C_{70}$ film on silicon substrate.

Assignment of the mode symmetries for $C_{70}$ is shown in Table 8.5. The table lists two assignments, one from the very early work on the subject in 1991 and the other is from one of the latest reports in 2008. It is interesting to note the slight difference in the results. In general, it can be noticed that modes below approximately 900 cm$^{-1}$ tend to have predominantly radial displacements, while the higher wavenumber modes have predominantly tangential displacements.

**Table 8.5** Symmetry assignment and frequency of Raman active modes in $C_{70}$. Early and current state of knowledge

| Meilunas et al. (1991) | | Wang & Fang ( 2008) | |
|---|---|---|---|
| $V_{RAMAN}$ | Mode | $V_{RAMAN}$ | Mode |
| 224(S) | | | |
| | | 227(s) | $E_2'$ |
| 229 5 | | | |
| 252(S) | | 253(S) | $A_1'$ |
| 260 | $A_1'$ | 259(S) | $A_1'$ |
| | | 303(m) | $E_2'$ |
| | | 309(S) | $E_2'$ |
| | | 327(m) | $A_1'$ |
| | | 361(w) | $A_1'$ |
| | | 382(w) | $A_1'$ |
| 400 | | 396(m) | $A_1'$ |
| | | 410(m) | $E_1''$ |
| 413 | | | |
| | | 419(w) | $E_1''$ |
| | | 431(m) | $E_2''$ |
| 436 | | | |
| 457 | | 457(S) | $A_1'$ |
| | | 481(w) | $E_1''$ |
| | | 490(vw) | $E_2''$ |
| 508 | | 509(w) | $E_1''$ |
| 521 | | 520(w) | $E_2'$ |
| | | 535(vw) | $E_2''$ |
| | | 568(S) | $A_1'$ |

| Meilunas et al. (1991) | | Wang & Fang ( 2008) | |
|---|---|---|---|
| $V_{RAMAN}$ | Mode | $V_{RAMAN}$ | Mode |
| 572 | $A_1'$ | | |
| | | 578(vw) | $A_1'$ |
| | | 640(vw) | $E_2'$ |
| | | 675(m) | $A_1'$ |
| | | 700(S) | $E_1''$ |
| 704 | $E_1'$ | | |
| 715 | | | |
| | | 721(m) | $A_1'$ |
| | | 737(S) | $E_1''$ |
| 740 | $E_1''$ or $E_2'$ | | |
| | | 796(m) | $E_2'$ |
| 771 | $E_1''$ or $E_2'$ | 801(m) | $E_2'$ |
| | | 898(m) | $E_2'$ |
| | | 947(vw) | $E_2'$ |
| | | 1012(w) | $E_1''$ |
| 1053(S) | | | |
| | | 1061(S) | $A_1'$ |
| 1063 | $A_1'$ | | |
| | | 1086(w) | $E_1''$ |
| 1167 | $A_2''$ | | |
| | | 1182(S) | $E_1''$ |
| 1187 | $A_1'$ | | $E_1''$ |
| | | 1228(S) | $E_2'$ |

*(Continued)*

**Table 8.5** (*Continued*)

| Meilunas et al. (1991) | | Wang & Fang ( 2008) | |
|---|---|---|---|
| $V_{RAMAN}$ | Mode | $V_{RAMAN}$ | Mode |
| 1232 | $A_1'$ | | |
| | | 1256(w) | $E_1''$ |
| 1259 | $E_1''$ or $E_2'$ | | |
| | | 1294(m) | $E_1''$ |
| 1301 | | | |
| 1316 | $E_1''$ or $E_2'$ | | |
| | | 1332(m) | $E_2'$ |
| 1335 | $E_1''$ or $E_2'$ | | |
| | | 1349(m) | $A_1'$ |
| | | 1367(m) | $A_1'$ |
| 1371 | $E_1'$ | | |
| | | 1373(w) | |
| 1439(S) | $E_1'$ | | $E_1''$ |
| 1443(S) | | | |
| | | 1445(S) | $E_2'$ |
| 1449 | $A_1'$ | | |
| 1461 (S) | | | |
| 1463(S) | | | |
| | | 1469(S) | $E_1''$ |
| 1471 | $A_1'$ | | |
| | | 1512(S) | $E_1''$ |
| 1515 | $E_1''$ or $E_2'$ | | |
| | | 1565(S) | $E_2'$ |
| 1569 | $A_1'$ | | |

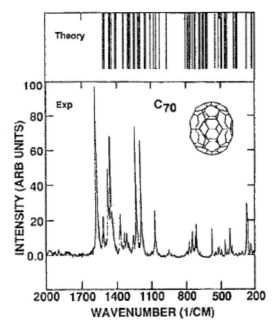

**Figure 8.29**  A nonpolarized Raman spectrum of $C_{70}$ film deposited on silicon substrate. Superimposed are the calculated Raman active modes. Reproduced with kind permission from Meilunas et al., *J. Appl. Phys.*, 1991, 70, 5128. Copyright American Institute of Physics, 1991.

Regarding Raman scattering of higher fullerenes, it can be said that very little is reported about the details of their vibrational spectra. Several factors contributed to this state of little knowledge including; lower symmetry, larger number of degrees of freedom and the many possible isomers of such higher fullerenes. More importantly, the lack of adequate quantities of well separated and characterized samples of higher fullerenes was a major reason for the lack of their experimental investigations.

### 8.9.2   Fullerene Solubility and Solvent Interactions

Solubility is perhaps the most important property of a substance to investigate if chemical performance of the substance is to be understood. In 1912, D. Tyrer reviewed the solubility theory as understood then and stated that, "The maximum amount of a substance that a liquid will hold in homogeneous solution at a given temperature, or as it is termed, the solubility of that substance in the given liquid, represents a physical constant which, *as yet*, it has been found impossible to connect in any consistent manner with any other physical constants or properties of either the solvent or the solute." In spite of the fact that during the 90 years followed, our understanding of "solubility" advanced dramatically, it can be said that we are still on a steep learning curve once fullerenes, and nanomoieties in general, are considered. In the early 1990s fullerene solubility in variety of solvents has been extensively investigated. Of all fullerene family members, $C_{60}$, $C_{70}$, solubilities were most investigated and reported in the literature. The results, in general, support the age-old principle *similia similibus solvuntur*, Latin for "like dissolves like." Table 8.6 shows the solubility of $C_{60}$ and $C_{70}$ in various solvents. It is interesting to note the wide range of solubility limits in certain solvents. Perhaps, the most important point to emphasize regarding the data presented in Table 8.6 is the huge reported difference in solubility limit between solvents that are not essentially different in their chemical structure or properties. For example, solubility limit of $C_{60}$ was reported to be 4.702 mg/ml and 17.928 mg/ml in 1,2,3-trimethylbenzene, and 1,2,4-trimethylbenzene, respectively. In addition, attempts to find a quantitative relation between the solubility limit (in molar fractions or in mg/ml) and traditional characteristics of solvents such as polarizability, polarity, molar volume, or Hildebrand's solubility parameter ($\delta$) have not been very successful. Another important point to note is the fact that $C_{60}$ is barely soluble in dimethylformamide (DMF) which is one of the best known solvents for SWCNT (a cylindrical version of fullerenes as we will discuss next chapter). This emphasizes the molecular shape and entropic effects on fullerene/solvent interactions. These and many other interesting phenomena observed for fullerene behavior in solutions added richness to the subject and sparked an interest that is still active and is expected to remain active for some time.

**Table 8.6**  Solubility limits of $C_{60}$ and $C_{70}$ in different solvents

| Solvent | Solubility of $C_{60}$ [mg/ml] | Solubility of $C_{70}$ [mg/ml] | References |
|---|---|---|---|
| | 1.699 | 1.116[c] | 280 |
| | 1.7 | | 290 |
| | 1.498 | | 281 |
| Benzene | 1.440 | | 283 |
| | 0.878 | | 285 |
| | 1.397 | | 285[b] |
| | 1.858 | | 286 |
| | 2.801 | | 280 |
| | 2.8 | | 290 |
| | 2.902 | | 281 |
| | 2.290 | 0.92884 | 291 |
| Toluene | 2.153 | 1.2024[c] | 283 |
| | 2.268 | | 287 |
| | 2.902 | 1.224 | 278 |
| | 2.398 | | 285[b] |
| | 3.197 | | 286 |
| | 8.712 | | 281 |
| 1,2-Dimethylbenzene | 7.344 | | 288 |
| | 9.288 | 13.392 | 278 |
| 1,3-Dimethylebenzene | 1.397 | | 281 |
| | 2.8290 | | 288 |
| 1,4-Dimethylebenzene | 5.897 | 3.4128[c] | 281 |
| | 3.139 | | 288 |
| 1,2,3- Trimethylebenzene | 4.702 | | 281 |
| 1,2,4-Trimethylebenzene | 17.928 | | 281 |
| | 1.498 | 1.26[c] | 280 |
| 1,3,5-Trimethylebenzene | 0.994 | | 283 |
| | 1.699 | | 281 |
| 1,2,3,4-Tetramethylebenzene | 5.803 | | 281 |

*(Continued)*

**Table 8.6** (Continued)

| Solvent | Solubility of $C_{60}$ [mg/ml] | Solubility of $C_{70}$ [mg/ml] | References |
|---|---|---|---|
| 1,2,3,5-Tetramethylebenzene | 20.880 | | 281 |
| Tetralin | 15.984 | | 280 |
| | 288.616 | | 289 |
| | 15.696 | | 285[b] |
| Ethylebenzene | 2.599 | | 281 |
| | 2.160 | | 288 |
| n-Propylbenzene | 1.498 | | 281 |
| Isopropylebenzene | 1.202 | | 281 |
| n-Butylbenzene | 1.901 | | 281 |
| sec-Butylbenzene | 1.102 | | 281 |
| Fluorobenzene | 0.590 | | 280 |
| | 1.202 | | 281 |
| Chlorobenzene | 6.998 | | 280 |
| | 5.702 | | 281 |
| Bromobenzene | 3.298 | | 280 |
| | 2.801 | | 281 |
| Iodobenzene | 2.102 | | 281 |
| 1,2-Dichlorobenzene | 27.000 | | 280 |
| | 24.624 | 31.032[c] | 281 |
| | 22.896 | 22.32 | 289 |
| | 23.400 | 25.704 | 292 |
| 1,2-Dibromobenzene | 13.824 | | 281 |
| 1,3-Dichlorobenzene | 2.398 | | 281 |
| | 5.033 | 16.056 | 292 |
| 1,3-Dichlorobenzene | 13.824 | | 281 |
| 1,2,4-Trichlorobenzene | 8.496 | | 280 |
| | 10.368 | | 281 |
| | 4.846 | | 293 |
| | 21.312 | | 286 |

| Solvent | Solubility of $C_{60}$ [mg/ml] | Solubility of $C_{70}$ [mg/ml] | References |
|---|---|---|---|
| Styrene | 3.751 | | 293 |
| *o*-Cresol | 0.0288 | | 280 |
| Benzonitrile | 0.410 | | 280 |
| Nitrobenzene | 0.799 | | 280 |
| Anisole | 5.602 | | 280 |
| | 6.696 | | 285[b] |
| *p*-Bromoanisole | 16.776 | | 285[b] |
| *m*-Bromoanisole | 16.200 | | 285[b] |
| Benzaldehyde | 0.389 | | 288 |
| Phenyle isocyanate | 2.441 | | 288 |
| Thiophenol | 6.912 | | 293 |
| 1-Methyl 2-nitrobenzene | 2.434 | | 288 |
| 1-Methyl 3-nitrobenzene | 2.362 | | 288 |
| Benzyle chloride | 2.398 | | 288 |
| Benzyle bromide | 4.939 | | 288 |
| 1,1,1-Trichloromethylbenzene | 4.802 | | 288 |
| 1-Methylenaphthalane | 32.976 | | 280 |
| | 33.192 | | 281 |
| 1-phenylnaphthalcne | 49.968 | | 280 |
| 1-Chloronaphthalene | 50.976 | | 280 |
| 1-Bromo,2-methylnaphthalene | 34.776 | | 281 |
| Xylenes | 5.2 | | 280, 290 |
| Mesitylene | 1.5 | | 280, 290 |
| Tetralin | 16.00 | | 280, 290 |
| **Alkanes** | | | |
| *n*-Pentane | 0.005 | 0.001728[c] | 280 |
| | 0.005 | | 290 |
| | 0.004 | | 283 |
| | 0.003 | | 294 |
| | 0.007 | | 286 |

(*Continued*)

**Table 8.6** (*Continued*)

| Solvent | Solubility of $C_{60}$ [mg/ml] | Solubility of $C_{70}$ [mg/ml] | References |
|---|---|---|---|
| | 0.043 | 0.0108[c] | 280 |
| | 0.040 | | 283 |
| *n*-Hexane | 0.037 | | 294 |
| | 0.052 | | 287 |
| | 0.046 | | 286 |
| 2-Methylepentane | 0.019 | | 294 |
| 3-Methylepentane | 0.025 | | 294 |
| *n*-Heptane | 0.048 | 0.04032[c] | 294 |
| | 0.2902 | | 285[b] |
| | 0.025 | 0.036[c] | 283 |
| | 0.025[b] | | 280, 290 |
| *n*-Octane | 0.020 | | 294 |
| | 0.2902 | | 285[b] |
| | 0.025 | | 286 |
| | 0.026 | | 283 |
| Isooctane | 0.026[b] | | 280, 290 |
| | 0.028 | | 286 |
| *n*-Nonane | 0.062 | | 294 |
| | 0.034 | | 286 |
| | 0.071 | 0.04536[c] | 280 |
| | 0.070[b] | | 280, 290 |
| *n*-Decane | 0.070 | | 283 |
| | 0.072 | | 286 |
| | 0.091 | 0.0864[c] | 283 |
| Dodecane | 0.091[b] | | 280, 290 |
| | 0.103 | | 286 |
| | 0.126 | | 283 |
| Tetradecane | 0126[b] | | 280, 290 |
| | 0.168 | | 286 |

| Solvent | Solubility of $C_{60}$ [mg/ml] | Solubility of $C_{70}$ [mg/ml] | References |
|---|---|---|---|
| **Cyclic Alkanes** | | | |
| Cyclopentane | 0.002 | | 280 |
| | 0.036 | 0.0684[c] | 280 |
| | 0.036[b] | | 290 |
| | 0.051 | | 283 |
| Cyclohexane | 0.036 | | 288 |
| | 0.036 | | 287 |
| | 0.035 | | 295 |
| | 0.054 | | 286 |
| Cyclohexene | 1.210 | | 293 |
| 1-Methyl,1-cyclohexene | 1.0290 | | 293 |
| Methylcyclohexane | 17.280 | | 293 |
| 1,2-Dimethylcyclohexane, mixture of *cis* and *trans* | 0.1290 | | 293 |
| Ethylcyclohexane | 0.252 | | 293 |
| 3:7 Mixture of *cis* and *trans* decalins | 4.601 | | 280 |
| | 1.872 | | 293 |
| *cis*-Decalin | 2.232 | | 280 |
| *trans*-Decalin | 1.296 | | 280 |
| **Haloalkanes** | | | |
| | 0.259 | 0.0684[c] | 280 |
| Dichloromethane | 0.252 | | 283 |
| | 0.2290 | | 288 |
| | 0.158 | | 280 |
| Trichloromethane | 0.173 | | 293 |
| | 0.511 | | 286 |
| | 0.317 | | 280 |
| Tertrachloromethane | 0.446 | | 283 |
| | 0.101 | | 296[a] |
| Dibromomethane | 0.360 | | 288 |

(*Continued*)

**Table 8.6** (*Continued*)

| Solvent | Solubility of $C_{60}$ [mg/ml] | Solubility of $C_{70}$ [mg/ml] | References |
|---|---|---|---|
| Tribromomethane | 5.638 | | 288 |
| Iodomethane | 0.770 | | 288 |
| Diiodomethane | 0.122 | | 293 |
| Bromochloromethane | 0.749 | | 293 |
| Bromoethane | 0.072 | | 293 |
| Iodoethane | 0.281 | | 280 |
| Trichloroethylene | 1.397 | | 280 |
| | 1.4 | | 290 |
| Tetrachloroethylene | 1.202 | | 280 |
| | 1.2 | | 290 |
| Dichlorodifloroethane | 0.022 | | 280 |
| 1,1,2-Trichlorotrifloroethane | 0.0288 | | 280 |
| 1,1,2,2-Tetrachloethane | 5.299 | | 288 |
| | 5.3 | | 280, 290 |
| 1,2-Dibromoethylene | 1.879 | | 288 |
| 1,2-Dichloroethane | 0.086 | | 280 |
| 1,2-Dibromoethane | 0.497 | | 288 |
| | 0.540 | | 288 |
| 1,1,1-Trichloroethane | 0.151 | | 293 |
| 1-Chloropropane | 0.022 | | 288 |
| 1-Bromopropane | 0.058 | | 288 |
| 1-Iodopropane | 0.2940 | | 288 |
| 2-Iodopropane | 0.122 | | 288 |
| 1,2-Dichloropropane | 0.101 | | 288 |
| 1,3-Dichloropropane | 0.360 | | 288 |
| 1,2-Dibromopropane | 0.403 | | 293 |
| 1,3-Diiodopropane | 2.765 | | 288 |
| 1,2,3-Trichloropropane | 0.778 | | 288 |
| 1,2,3-Tribromopropane | 7.013 | | 288 |

| Solvent | Solubility of $C_{60}$ [mg/ml] | Solubility of $C_{70}$ [mg/ml] | References |
|---|---|---|---|
| 1-Chloro-2-methylpropane | 0.029 | | 288 |
| 1-Bromo-2-methylpropane | 0.086 | | 293 |
| 1-Iodo-2-methylepropane | 0.338 | | 288 |
| 2-Chloro-2-methylepropane | 0.010 | | 293 |
| 2-Bromo-2-methylpropane | 0.060 | | 293 |
| 2-Iodo-2-methylepropane | 0.2290 | | 293 |
| Bromobutane | 1.202 | | 285[b] |
| Cyclopentyl bromide | 0.410 | | 288 |
| Cyclohexyle chloride | 0.533 | | 288 |
| Cyclohexyle bromide | 2.203 | | 288 |
| Cyclohexyle iodide | 8.064 | | 288 |
| Bromoheptane | 2.297 | | 285[b] |
| Bromooctane | 3.398 | | 285[b] |
| 1-Bromotetradecane | 6.192 | | 285[b] |
| 1-Bromooctadecane | 6.192 | | 285[b] |
| Methylene chloride[b] | 0.254 | | 280, 290 |
| **Alcohols** | | | |
| Methanol | 0.000 | | 294 |
| Ethanol | 0.001 | | 280 |
| | 0.001 | | 294 |
| 1-Propanol | 0.004 | | 294 |
| 1-Butanol | 0.009 | | 294 |
| 1-Pentanol | 0.0290 | | 294 |
| 1-Hexanol | 0.042 | | 294 |
| 1-Octanol | 0.047 | | 294 |
| 2-propanol | 0.002 | | 294 |
| 2-Butanol | 0.004 | | 294 |
| 2-Pentanol | 0.0294 | | 294 |
| 3-Pentanol | 0.029 | | 294 |
| 1,3-Propandiol | 0.001 | | 294 |

(*Continued*)

**Table 8.6** (*Continued*)

| Solvent | Solubility of $C_{60}$ [mg/ml] | Solubility of $C_{70}$ [mg/ml] | References |
|---|---|---|---|
| 1,4-Butandiol | 0.002 | | 294 |
| 1,5-Pentadiol | 0.004 | | 294 |
| *n*-Methyle-2-pyrrolidone | 0.89 | | |
| **Naphthalenes** | | | |
| 1-Methylnaphthalene | 33.0 | | 280, 290 |
| Dimethylnaphthalene | 36.0 | | 280, 290 |
| 1-Phenylnaphthalene | 50.0 | | 280, 290 |
| 1-Chloronaphthalene | 51.0 | | 280, 290 |
| **Other Polar Solvents** | | | |
| Nitromethane | 0.000 | | 280 |
| | 0.216 | | 285[b] |
| Nitroethane | 0.002 | | 280 |
| Acetone | 0.001 | 0.001656[c] | 280 |
| Acetonitrile | 0.000 | | 280 |
| Acrylonitrile | 0.004 | | 293 |
| *n*-Butylamine | 3.686 | | 293 |
| 2-Methyloxyethyl ether | 0.032 | | 288 |
| *N,N*-dimethylflormamide | 0.027 | | 288 |
| Dioxane | 0.041 | | 283 |
| Water | 0.000 | 1.15E-10 | 294 |
| **Miscellaneous** | | | |
| Carbon disulfide | 7.920 | 8.496[c] | 280 |
| | 5.162 | 13.104 | 283 |
| | 7.704 | | 278 |
| | 7.488 | | 293 |
| | 11.808 | | 286 |
| Thiophene | 0.403 | | 281 |
| | 0.238 | | 293 |

| Solvent | Solubility of $C_{60}$ [mg/ml] | Solubility of $C_{70}$ [mg/ml] | References |
|---|---|---|---|
|  | 0.058 |  | 280 |
| Tetrahydrofuran | 0.576 |  | 285[b] |
|  | 0.360 |  | 286 |
| 2-Methylthiophene | 6.797 |  | 280 |
|  | 0.893 |  | 280 |
| Pyridine | 0.2902 |  | 281 |
|  | 0.2902 |  | 285[b] |
| Piperidine | 53.280 |  | 285[b] |
| 2,4,6-Trimethylpyridine | 8.712 |  | 285[b] |
| Pyrrolidine | 47.520 |  | 285[b] |
| N-Methyl-2-pyrrolidone | 0.893 |  | 280 |
| Tetrahydrothiophene | 0.0290 |  | 280 |
|  | 0.1288 |  | 293 |
| Quinoline | 7.200 |  | 281 |
| Isopropanol |  | 0.00294[c] |  |
| Carbon tetrachloride | 0.32 | 0.1008[c] | 280, 290 |

[a]Measurements taken at 298 K, [b]Measurements taken at 291 K, [c]Measurements taken at 303 K

### 8.9.3 Solvent Effects on Fullerenes

Many investigations observed the solvent effect on the fullerene properties as detected by optical absorption and nuclear magnetic resonance (NMR), however, very few investigations have reported on solvent effects on Raman frequencies of fullerenes. Even fewer investigations have been devoted to calculating the change in Raman frequencies as the fullerene molecule interacts with different solvating environments. Amer et al. utilized semi-empirical molecular simulation methods to predict the effect of water interaction on the Raman frequencies of $C_{60}$. Figure 8.30 shows the calculated Raman frequency shifts in the three Raman active surface modes [$A_g(2)$,

$H_g(7)$, and $H_g(8)$] of a $C_{60}$ molecule as a result of interaction with increasing number (from 1 to 64) of water molecules.

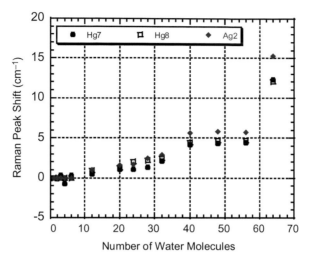

**Figure 8.30** Raman peak shifts of fullerene surface modes at different number of interacting water molecules. Reproduced with kind permission from Amer et al., *Chem. Phys. Lett.*, 2005, 411, 395. Copyright Elsevier, 2005.

As shown in Figure 8.30, at low number of interacting water molecules no significant shifts in the Raman frequencies is predicted. The frequency starts to shift at increasing rate as the number of interacting molecules increased until it reached about 40 water molecules. The Raman shift formed a plateau beyond this level before increasing again as the number of interacting molecules reached 64. The calculated changes in the Raman frequencies were explained as follows; at low number of interacting water molecules, and due to the hydrophobic nature of the fullerene molecule, water molecules will interact weakly with the fullerene, and hence, little or no change in the Raman frequencies of the fullerene molecule is predicted. As the number of water molecules increases to a certain limit, the water molecules around the fullerene form an intermolecular hydrogen-bonded network that stabilizes the interaction with the water causing the sharp increase in the Raman peak position predicted at 40 interacting water molecules. Addition of more water molecules beyond this limit will not lead to increasing interaction. The plateau in peak position can be rationalized in terms of water molecules that are not directly interacting with the fullerene which is shielded by the water cage already formed around it. As the number of water molecules is further increased to the limit that an additional tight-bonded cage can be formed, thereby, further increasing the blue shift in Raman peak position.

This study drags attention to a very important point; the mutual water/water and water/fullerene interactions not only affect the physicochemical properties of the fullerene molecules (as reflected in their NMR, optical absorption properties, and Raman characteristics) but also affect the solvent structure. Figure 8.31 shows the calculated equilibrium structure of a fullerene molecule interacting with 12 water molecules. It is clear from the figure that the mutual and simultaneous interaction between the water and fullerene molecules resulted in the shown thread-like structure not common in pure water.

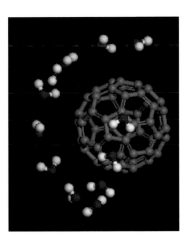

**Figure 8.31** Simulated structure of a fullerene molecule interacting with 12 water molecules. Reproduced with kind permission from Amer et al., *Chem. Phys. Lett.*, 2005, 411, 395. Copyright Elsevier, 2005.

More recent investigations further elucidate the important and unique role played by the solvent in the formation of versatile fullerene–based meso-systems and devices with unique structures and, hence, properties and performance. For example, it was shown that the performance of fullerene-based solar cells is largely affected by the fullerene structure which, in turn, is controlled by the solvent used in the processing.

In addition, solvent geometry has critical effect on the determination of [60] fullerene self-assembly into dot, wire and disk structures via a droplet drying process. Figure 8.32 schematically shows the solvent structure and the self-assembled fullerene structures resulting upon evaporation.

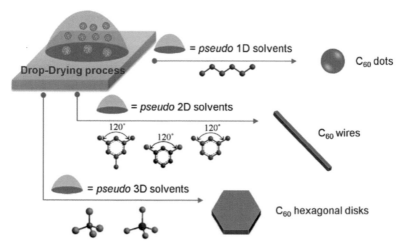

**Figure 8.32** The self-assembly of dimensionally-confined $C_{60}$ structures via a solution drop-drying process. Systematic correlations in the formation of *pseudo* 2-D (hexagonal disk), *pseudo* 1-D (wire) and *pseudo* 0-D (dot) $C_{60}$ structures from $C_{60}$ solutions of $p3D$, $p2D$, and $p1D$ solvents, respectively. Reproduced with kind permission from Park et al., *Chem. Commun.*, 2009, 2009, 4803. Copyright Royal Society of Chemistry, 2009.

The study shows that the morphologies of $C_{60}$ crystalline structures formed via evaporation are precisely determined by the molecular geometries of solvents. It experimentally establishes the specific correlation of solvent molecular geometry to directed assembly of $C_{60}$ structure. The study demonstrates that while solvents with *pseudo* 3-D ($p3D$) molecules direct $C_{60}$ molecules to assemble into *pseudo* 2-D hexagonal disk structures, solvents with $p2D$ and $p1D$ molecular structures result in $p1D$ $C_{60}$ (wires) and $p0D$ $C_{60}$ spheroidal structures, respectively. Figure 8.33 shows scanning electron microscope (SEM) and optical microscope images of $C_{60}$ crystalline structures obtained from $C_{60}$ solutions in (a) $p3D$ $CCl_4$, (b) $p2D$ m-xylene, and (c) $p1D$ hexane.

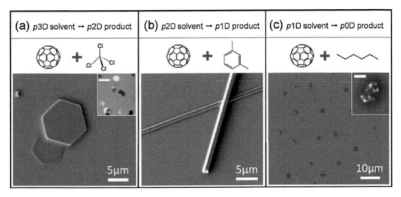

**Figure 8.33** SEM and optical microscope images of $C_{60}$ crystalline structures obtained from $C_{60}$ solutions in (a) $p$3D CCl$_4$, (b) $p$2D m-xylene, and (c) $p$1D hexane. The scale bars of the insets in (a) and (c) are 30 mm and 500 nm, respectively. Reproduced with kind permission from Park et al., *Chem. Commun.*, 2009, 2009, 4803. Copyright Royal Society of Chemistry, 2009.

In spite of such a major leap in correlation finding, the details of the mechanism responsible for such correlation are still to be understood. Understanding the physics of such correlations is essential for our ability to further develop nanotechnology. In addition to its importance in enabling the development of technologically advanced nanostructured devices with unique properties, the ability to understand the correlation between the environment and self-assembly or directed assembly of structures would, definitely, have a major impact on our ability to understand the mechanisms of many biological processes involving growth and coagulation. A crucial point to realize in the aforementioned study is that the geometrical shape of the solvent molecules was the major factor in determining the structure of the directed assembly. This emphasizes the critical role played by molecular geometry that is related to entropic effects in the assembly of nanosystems. Molecular geometry as related to entropic effects on the directed assembly of nanostructures is a major frontier that is currently being explored. Knowledge generated in colloidal science investigations should be a solid base to build upon.

Another interesting issue to point out is the finding that while free evaporation of $n$-dimensional solvents containing $C_{60}$ molecules led to the assembly of an $(n-1)$-dimensional $C_{60}$ structures, coagulation of $C_{60}$ saturated toluene (2-D solvent) solution into alcohols (1-D liquid) (*via* liquid-liquid interface precipitation LLIP technique) led to the assembly of $C_{60}$ molecules into a 1-D (wire, whiskers, or rod) structures as shown in Figure 8.34. The length of the assembled 1-D structures was found to depend on the molecular *length* of the alcohol. Coagulation in methanol, ethanol, and propanol led to the formation of 1-D assemblies with average length shortest for propanol and longest for methanol. Figure 8.35 shows the length distribution for $C_{60}$ wires directly assembled by LLIP technique in three different alcohols. It is clear from the figure that while 95% of the wires directly assembled by methanol have length less than 150 μm, 95% of the wires assembled in ethanol and propanol have length less than 100 μm and 50 μm, respectively.

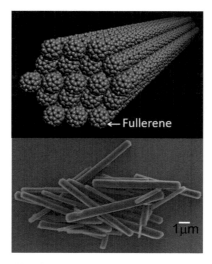

**Figure 8.34** Assembly of [60] fullerene into one-dimensional fullerene micro-whiskers. Reproduced with kind permission from Miyazawa et al., in *Fullerene Nanowhiskers* (2nd Ed.), Copyright Jenny Stanford Publishing, 2019.

### 8.9.4 Fullerene Effects on Solvents

When a fullerene interacts with a solvent, not only the solvent affects the fullerene properties and behavior as we discussed in the previous section, but it was reported that the fullerene molecule also affects the solvent. Mainly, it was reported that a [60] fullerene molecule affects the structure of aromatic solvents it interacts with. In fact, the high solubility of $C_{60}$ in benzene, toluene, para-xylene, and other aromatic solvents, itself, indicates strong interaction, and hence, suggests that the structure of the aromatic solvent can change in response to the presence of the dissolved fullerene. Structuring of aromatic solvents interacting with $C_{60}$ was assumed based on observed anomalies in many of the physical properties and behavior of the solvent. Nonmonotonic changes in the solution density were observed with monotonic increase in the concentration of fullerene molecules in the solution. Their small- and wide-angle X-ray diffraction and permittivity experimental results pointed out the possibility of solvent structuring around the spheroidal fullerene. Figure 8.36 shows wide-angle X-ray diffraction scattering curves ($Cu_{K\alpha}$ radiation) of fullerene $C_{60}$ solutions in (a) *p*-xylene, (b) toluene, and (c) benzene with different fullerene concentration.

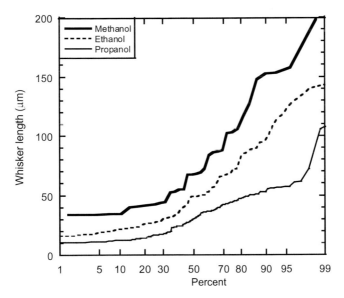

**Figure 8.35** Length distribution for $C_{60}$ wires directly assembled by LLIP technique in three different alcohols (Image created by Prof. M. S. Amer, 2020).

Brillouin scattering measurements of adiabatic compressibility and evaporation kinetic measurements on $C_{60}$/toluene solutions of different concentrations could also be interpreted along the notion of solvent structuring around the fullerene molecule. Figure 8.37 shows the adiabatic compressibility of $C_{60}$/toluene solutions as measured using Brillouin spectroscopy. The observed linear increase in the solution compressibility is indicative of solvent structuring rendering an open structure with higher compressibility.

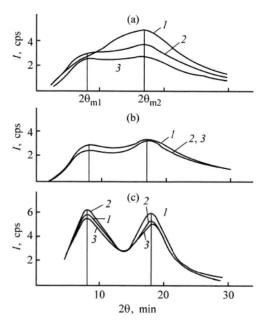

**Figure 8.36** Wide-angle X-ray diffraction scattering curves ($Cu_{K\alpha}$ radiation) of fullerene C60 solutions in (a) *p*-xylene, (b) toluene, and (c) benzene. Fullerene concentration, %: (a) (*1*) 0 (pure *p*-xylene), (*2*) 0.005, and (*3*) 0.05; (b) (*1*) 0 (pure toluene), (*2*) 0.005, and (*3*) 0.05; and (c) (*1*) 0 (pure benzene), (*2*) 0.001, and (*3*) 0.075. Reproduced with kind permission from Ginzburg and Tuichiev, *Crystallogr. Rep.*, 2008, 53, 645. Copyright Pleiades Publishing, 2008.

Recent molecular dynamic simulations showed that once the [60] fullerene molecule is introduced in the aromatic, the presence of $C_{60}$ and its strong interaction with solvent molecules causes a sort of molecular structuring in the solvent leading to the formation of a tightly bound toluene shell around the nanospheres (about 1 nm thick). Such a $C_{60}$/solvent clathrate would be covered by a solvo-phobic (open structure) shell. The thickness of the outer radius of the solvo-phobic shell was estimated to be 6.6 nm (for toluene) and differs depending on the solvent nature. This means that *the long-range interaction distance* around the fullerene molecule in toluene is around 6.6 nm. Such structuring interpretation of the observed experimental results was confirmed using molecular dynamic simulations. Figure 8.38 shows the equilibrium structure of $C_{60}$ interacting with chlorobenzene solvent as calculated using molecular dynamic simulations. It is important to note the structuring and the open areas in the solvent.

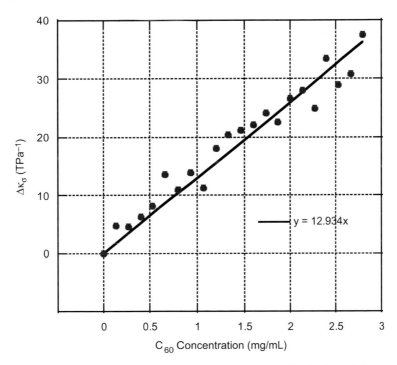

**Figure 8.37** The change in adiabatic compressibility ($\Delta \kappa_s$) of $C_{60}$/toluene solution as a function of $C_{60}$ concentration. Reproduced with kind permission from Amer et al., *Chem. Phys. Lett.*, 2008, 457, 329. Copyright Elsevier, 2008.

### 8.9.5  Mechanical Properties of Spherical Fullerenes

The one spherical fullerene molecule that was most investigated from mechanical properties viewpoint is the $C_{60}$ molecule. Due to the impossibility of experimental measurements, most of the investigations were done using computer simulation techniques. Molecular dynamic simulations investigating the compressibility of individual $C_{60}$ molecules showed that their bulk modulus is around 1.1 TPa. The bulk modulus was also reported to increase linearly with applied hydrostatic pressure reaching a value of 1.4 TPa at applied pressure of 10 GPa. Most interestingly, studies showed that interaction with water and methanol, as solvents, significantly affect the fullerene pressure-volume thermodynamic equation of state (EOS) leading to significantly less compressible fullerene molecule. Figure 8.39a shows the simulated $P$–$V$ (EOS) for an isolated $C_{60}$ molecule not interacting with a solvent and those interacting with water and methanol, separately, and Figure 8.39b shows the fullerene bulk modulus calculated based on the simulated EOS. The results drag the attention to the important point that mechanical "constants" of fullerene, as a nanosystem, cannot be considered out of their chemical environment context.

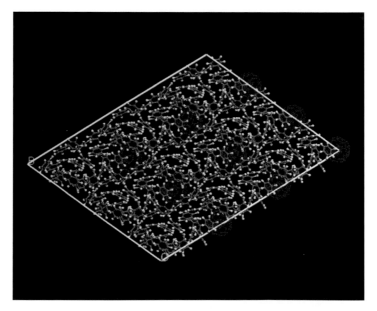

**Figure 8.38** Equilibrium structure of $C_{60}$ interacting with chlorobenzene as obtained using molecular dynamic calculations. Note the solvent structuring and the open areas within the solvent. Reproduced with kind permission from Amer et al., *Phys. Chem. Chem. Phys.*, 2018, 20(16), 11296–11305. Copyright RSC, 2018.

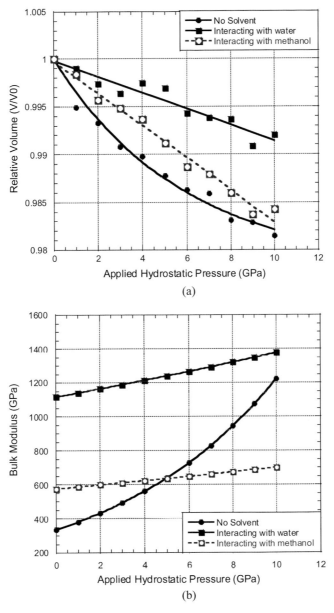

**Figure 8.39** Molecular dynamic simulation results for (a) *P–V* equation of state calculated for single fullerene molecules interacting with different chemical moieties, and (b) calculated values for the fullerene bulk modulus based on the simulated EOS. Reproduced with kind permission from Amer and Maguire, *Chem. Phys. Lett.*, 2009, 476, 232. Copyright Elsevier, 2009.

### 8.9.6 Spherical Fullerene-Based 2-D Materials

2-D materials are the state-of-the-art field in materials science and engineering. This class of materials started with the successful production of graphene in 2004 and extended to many other materials due to the very unique set of properties they exhibit. 2-D materials are defined as sheet materials with only one unit cell thickness. 2-D films based on a single layer of $C_{60}$ molecules were recently produced, using Langmuir–Blodgett (LB) technique, and investigated. While the film is a monolayer thick, it is made of hexagonal closed packed (hcp) $C_{60}$ molecules within the film plane as shown in Figure 8.40. The figure shows a $8 \times 8$ nm scanning tunneling microscope (STM) picture of the film surface. Note the hcp arrangement of the fullerene molecules within the film. Figure 8.41 shows the LB isotherm for a hcp $C_{60}$ monolayer film. The LB isotherm is basically a 2-D load-deformation curve similar to those acquired in axial tensile or compression testing machine. From such isotherms, the film compressive stress-strain curve can be obtained and film stiffness can be determined. Figure 8.42 shows the compressive stress-strain curve for a monolayer hcp film of $C_{60}$ molecules. The linear elastic behavior of the film is very clear. The film stiffness was measured to be 32.1 MPa. It is important to note that while the individual spherical fullerene has extremely high stiffness, the film exhibit a much lower stiffness since the molecules are held together by weak intermolecular forces and mechanical friction.

Interestingly enough, the 2-D film stiffness showed dependence on the nature of solvent used to process the film. Figure 8.43 shows the $C_{60}$-based 2-D film stiffness plotted against the solubility limit of the $C_{60}$ in the solvent used to produce the film. It is clear from the figure that a maximum film stiffness (around 50 MPa) is reached around solubility limit of 5.2 mg/ml (for xylene). Solvents with increased solubility limits (chlorobenzene) resulted in films with lower stiffness. These results are another clear example for the important effect that molecular shape, size, and polarity can play in nanostructured systems.

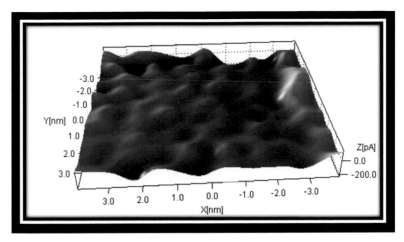

**Figure 8.40** STM image of 8 × 8 nm area of a monolayer $C_{60}$ film. Reproduced with kind permission from Amer and Al Talebi, *Mater. Res. Express*, 2018, 5(1), 016407. Copyrights IOP, 2018.

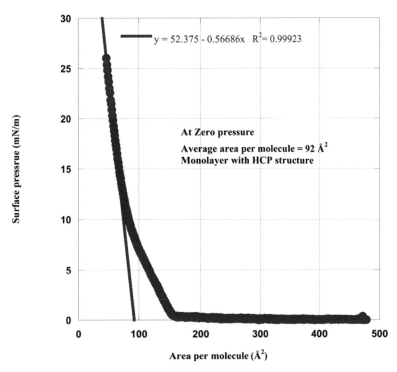

**Figure 8.41** LB isotherm for a monolayer $C_{60}$ film. Reproduced with kind permission from Amer and Al Talebi, *Mater. Res. Express*, 2018, 5(1), 016407. Copyrights IOP, 2018.

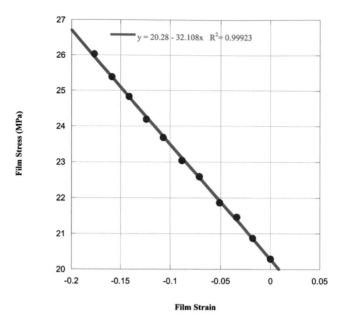

**Figure 8.42** Stress/strain curve for a hcp monolayer of $C_{60}$ under compressive stresses. Compressive film stiffness 32.1 MPa. Reproduced with kind permission from Amer and Al Talebi, *Mater. Res. Express*, 2018, 5(1), 016407. Copyright IOP, 2018.

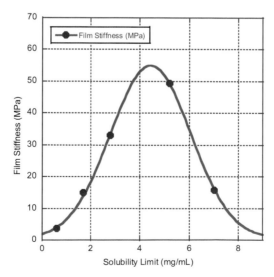

**Figure 8.43** Stiffness of hcp $C_{60}$ monolayer film processed using different aromatic solvents as a function of the fullerene solubility limit (Image created by Prof. M. S. Amer, 2020).

## Problems

1. How many hexagons and pentagons will exist in the following spherical fullerenes? $C_{60}$, $C_{80}$, and $C_{120}$.
2. What is the Euler's pentagon rule to construct stable closed-cage like molecules?
3. What is the smallest possible spherical fullerene?
4. 0-D fullerenes are typically expressed as $C_n$, what are the restrictions on "$n$"?
5. What is the symmetry group of the $C_{60}$ molecule?
6. How many isomers are predicted to be stable for the following 0-D fullerenes: $C_{60}$, $C_{70}$, and $C_{80}$?
7. Discuss any stability issues in 0-D fullerenes.
8. Which is the smallest fullerene molecule with no abutting pentagons?
9. Are all atoms of a $C_{60}$ fullerene correlated? Explain your answer.
10. Are all carbon atom sites in a $C_{60}$ equivalent?
11. Are all carbon atom sites in a $C_{70}$ equivalent?
12. Calculate the diameter of $C_{120}$ assuming the icosahedron symmetry.
13. State the spherical fullerenes production methods and discuss each method.
14. What are the processing parameters known to affect the fullerene yield in an arc-discharge process?
15. What are the sublimation temperatures for $C_{60}$ and $C_{70}$?
16. What is the advantage of extraction using the sublimation method?
17. What are the experimental techniques used to verify the effectiveness of a purification method?
18. Explain the liquid chromatography technique for fullerene purification?
19. How can you quantitatively identify different fullerenes? Give examples.
20. How many Raman active vibration modes a $C_{60}$ molecule exhibits?
21. Explain the difference between molecular vibration modes and crustal vibration modes.

22. What is the most stable crystal structure for $C_{60}$ molecules?

23. How many Raman active vibration modes a $C_{70}$ molecule exhibits?

24. Is there a quantitative relation between the solubility limit and traditional characteristics of solvents?

25. Would the solvent used to process fullerene-based solar cells affect the cell performance? Why?

26. Explain the effect of solvent molecular geometry on the geometry of self-assembled $C_{60}$ clusters.

27. The perturbation fields around a spherical fullerene affect its optical and mechanical properties. Discuss this statement and give examples.

28. Does the stiffness of 2-D monolayer films based on $C_{60}$ depends on the nature of solvent used during processing? Discuss your answer.

# Chapter 9

# One-Dimensional Fullerenes, Carbon Nanotubes

## 9.1    Introduction

One-dimensional (1-D) fullerenes or tubular fullerenes were the second type of fullerenes to be discovered and investigated. In this chapter, we will discuss their different types, structure, physical properties, and interaction with different types of solvents. In addition, production, and purification techniques will be discussed. The properties of 2-D monolayer films based on aligned nanotubes will be discussed as well.

## 9.2    Single-Walled Carbon Nanotubes

1-D fullerenes are typically in the form of tubes or cylinders with diameters ranging between a fraction of a nanometer and few tens of nanometers. They can be in the form of a single tube that we refer to as single-walled carbon nanotubes (SWCNTs), or in the form of a group of concentric tubes positioned one inside the other, and in this case, we refer to them as multi-walled carbon nanotubes (MWCNTs). The tubes can also be in the form of only two concentric tubes and in this case, they are referred to as double-walled carbon nanotubes (DWCNTs). Carbon nanotubes can be formed by rolling

*Gigantic Challenges, Nano Solutions: The Science and Engineering of Nanoscale Systems*
Maher S. Amer
Copyright © 2022 Jenny Stanford Publishing Pte. Ltd.
ISBN 978-981-4877-74-9 (Hardcover), 978-1-003-14704-6 (eBook)
www.jennystanford.com

graphene sheets (a single sheet, double sheets, or multiple sheets) into a seamless cylinder. A schematic of how graphene sheets can be rolled to form carbon nanotubes is shown in Figure 9.1. From the science and application viewpoint, SWCNTs are more interesting and exciting. Hence, we will start by describing their structure first.

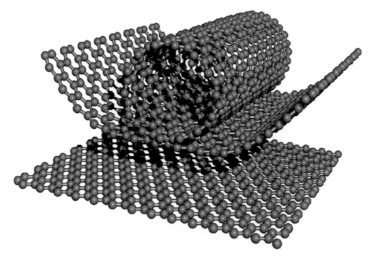

**Figure 9.1** A schematic of how graphene sheets can be rolled to form carbon nanotubes. Reproduced with kind permission from Endo et al., *Philos. Trans. R. Soc. London*, Ser. A, 2004, 362, 2223. Copyright The Royal Society of Chemistry, 2004.

A SWCNT can be formed by rolling a graphene sheet. The direction at which the sheet is rolled determines the structure and the properties of the resulting tube. Figure 9.2 shows the typical hexagonal (honeycomb) lattice structure of a graphene sheet made of covalently bonded $sp^2$ hybridized carbon atoms. The two principal axes directions and the unit vectors ($a_1$ and $a_2$) of the primitive hexagonal cell are shown in the figure. If we connect any two carbon atoms on the flat sheet geometry by a straight line, we can always roll the sheet around an axis normal to that line such that the selected line becomes a circumference of the tube. In this case, the two atoms at the start point and end point of the selected line has to coincide and become the same atom of the tube surface (see Figure 9.2). If we assign any random atom on the sheet the coordinates (0,0) and consider that point as the origin of a 2-D space, the coordinates of any other atom on the flat sheet can be

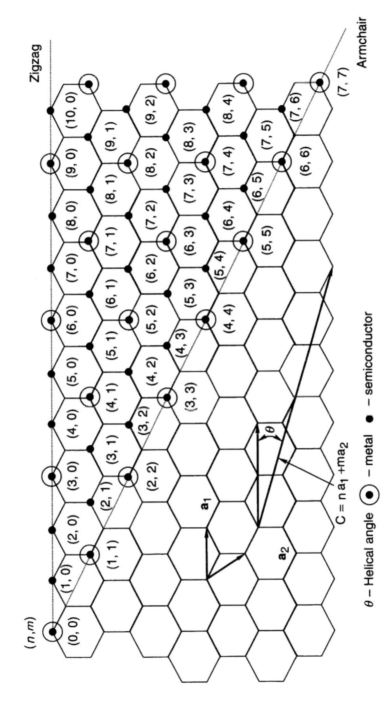

**Figure 9.2** Typical hexagonal structure of a graphene sheet made of covalently bonded sp² hybridized carbon atoms. Adapted with kind permission from Hu et al., *Acc. Chem. Res.*, 1999, 32, 435. Copyright Elsevier, 1999.

easily described in terms of the unit vectors ($a_1$ and $a_2$) as shown in Figure 9.2. Then, the line connecting any two carbon atoms on the flat sheet geometry can always be expressed as a vector ($C_h$) in terms of multiples of the unit vectors ($a_1$ and $a_2$). Assigning the vector start point the origin coordinates (0,0) would simplify the representation and the circumference of a carbon nanotubes can be expressed in vector notation as:

$$\overrightarrow{C_h} = n\overrightarrow{a_1} + m\overrightarrow{a_2} \equiv (n,m) \tag{9.1}$$

The vector representing the circumference of a tube ($p$) is the chiral vector ($C_h$). The angle between the chiral vector and the $a_1$ vector is known as the chiral angle ($\theta$) (see Figure 9.2). A SWCNT can be uniquely defined by its chiral vector. The word uniquely, in this context, means both geometrically, and as we will realize later, physical and chemical property wise as well. Hence nanotubes are usually referred to by the two vector multiples ($n,m$) defining their chiral.

Once the sheet is rolled to form a tube, the carbon atoms will form a helix of graphite lattice points will be formed. For each tube with a chiral angle ($\vartheta$) between 0° and 30° there is an equivalent tube with a chiral angle $\vartheta' = \vartheta + 30°$. The two equivalent tubes will have two equivalent right-hand and left-hand helices. Due to the six-fold rotation symmetry of the graphene lattice, chiral vectors that are multiple of 60° apart are equivalent and the tubes having such chiral vectors are essentially the same [a ($n,m$) and a ($m,n$) tubes are equivalent]. For these two reasons, chirality discussions of nanotubes in the literature are restricted to chiral angles $0 \leq \theta \leq 30°$. Figure 9.3 shows a scanning tunneling microscopy image of two SWCNTs depicting the helix and the chiral angle of the tube.

Depending on the chiral vector of the tube, or in other words, depending on the chiral angle of the tube, the architectural configuration of C–C bonds in SWCNTs can be classified into three different categories; armchair tubes (the term *serpentine* tubes might also be encountered in the literature), zigzag tubes (the term *sawtooth* tubes might also be encountered in the literature), and chiral tubes. As shown in Figure 9.2, armchair tubes are formed when the chiral vector of the tube results in a chiral angle that equals 30°. In this case, the two tube vector multiples are equal to each other, i.e., $n = m$. All zigzag tubes will have a chiral angle equal to zero

($\theta_{-}= 0$) and their *m vector* multiple will always equal zero (see Figures 9.2 and 9.3). Any other tube with chiral angle ($0 < \theta < 30°$) is classified as a chiral tube. Figure 9.4 shows the C–C bond arrangements in both zigzag and armchair tubes.

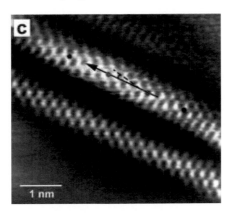

**Figure 9.3** STM images of SWNTs. The solid, black arrows highlight the tube axes, and the dashed lines indicate the zigzag directions. Reproduced with kind permission from Hu et al., *Acc. Chem. Res.*, 1999, 32, 435. Copyright American Chemical Society, 1999.

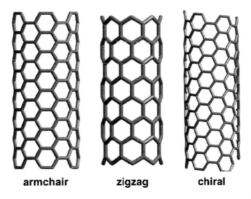

armchair          zigzag          chiral

**Figure 9.4** Configurations of C–C bonds in armchair, zigzag, and chiral single-walled carbon nanotubes.

Hence a (10,10) tube is an arm chair tube, and a (8,0) tube is a zigzag tube. Any tube with mixed vector multiples, for example (7,8) is a chiral tube. In order to describe these three different arrangements in SWCNTs we can say that; in armchair tubes, the hexagons are pointing normal to the tube axis, in zigzag tubes, the hexagons are pointing along the tube axis, and in chiral tubes, the hexagons are pointing at an angle to the tube axis (see Figure 9.4).

From Figure 9.2, it is clear that the unit vectors $a_1$ and $a_2$ have equal magnitude (length). This magnitude can be correlated to the C–C bond length in the graphene sheet. Taking the C–C bond length in the graphene sheet as 1.421 Å leads to:

$$|a_1| = |a_2| \equiv a = 1.421\sqrt{3} \text{ Å} = 2.461 \text{ Å} \qquad (9.2)$$

In addition, the length of the chiral vector ($C_h$), and hence, the tube circumference ($p$), can be calculated using the unit vector magnitude ($a$) and the tube vector multiples as:

$$p = |C_h| = a\sqrt{n^2 + nm + m^2} \qquad (9.3)$$

Hence, the diameter of a SWCNT ($d$) can be calculated using its chirality or vector multiples as:

$$d = \frac{a}{\pi}\sqrt{n^2 + nm + m^2} \qquad (9.4)$$

It is important to note that the calculated tube diameter according to Eq. 9.4 is, actually, a theoretical diameter calculated based upon pure geometrical considerations and an assumed length of the C–C bond. As we have mentioned during our discussion of 1-D fullerene structures, the C–C bond length is known to change due to surface curvature as well as due to the presence of chemical potentials close to the fullerene molecules. This, however, does not make Eq. 9.4, a less valuable tool in obtaining acceptable *estimations* for the diameters of a SWCNT.

According to Eq. 9.4, the diameter of SWCNTs can be readily estimated based on their vector multiples. For example, a (10,10) armchair nanotube would have a theoretical diameter of 1.37 Å. Also, a (9,0) zigzag nanotube should have a theoretical diameter of 7.15 Å. These values should also be increased roughly by 3.37 Å if the thickness of π-electron shells to be accounted for. This, again, puts SWCNTs in the center of interest as nanosystems by themselves and makes them ideal as building blocks for nanostructured systems.

The chiral angle ($\theta$) can also be defined using the tube chirality multiples as

$$\sin\theta = \frac{\sqrt{3}m}{2\sqrt{n^2 + nm + m^2}}; \cos\theta = \frac{2n + m}{2\sqrt{n^2 + nm + m^2}} \qquad (9.5)$$

It is important to note that the chirality of a SWCNT not only determines the tube's geometrical properties (such as diameter) but mostly its thermal, optical and electronic properties as well as we will discuss later in this chapter. Based on their chirality, the electronic band structure of nanotubes was found to exhibit either metallic or semiconducting behavior. It was shown that metallic conduction occurs when

$$n - m = 3q \tag{9.6}$$

where $q$ is an integer (0, 1, 2, 3, ...). This means that all armchair single-walled carbon tubes should be metallic, while one third of zigzag and chiral tubes are expected to be metallic. The reminder of SWCNTs is expected to be semiconducting.

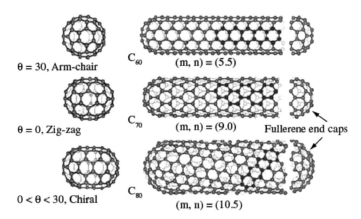

$\theta = 30$, Arm-chair    $C_{60}$    $(m, n) = (5.5)$

$\theta = 0$, Zig-zag    $C_{70}$    $(m, n) = (9.0)$    Fullerene end caps

$0 < \theta < 30$, Chiral    $C_{80}$

$(m, n) = (10.5)$

**Figure 9.5** Fullerene molecules end caps compatible with different types of single-walled carbon nanotubes. Reproduced with kind permission from Kalamkarov et al., *Int. J. Solids Structr.*, 2006, 43, 6832. Copyright Elsevier, 2006.

The ends of SWCNTs are typically capped with one half of a fullerene molecule end caps. The nature of the fullerene end cap depends on the diameter and chirality of the nanotube. Figure 9.5 shows different SWCNTs capped with different half-fullerene molecules. As shown in the figure, one half of a $C_{60}$ molecule can correctly cap each end of a (5,5) armchair tube. One half of a $C_{70}$ molecule can correctly cap the ends of a (9,0) zigzag tube, and one half of a $C_{80}$ fullerene molecule can correctly cap the ends of a (10,5) chiral nanotube. Due to the large aspect ratio of SWCNTs (>1000), the end caps can be neglected in tube analysis without losing the generality of the treatment. In addition, the same fullerene molecule can cap different types of SWCNTs. For example, the $C_{60}$ molecule can cap both the (5,5) armchair and the (9,0) zigzag nanotubes. Figure 9.6 shows the two types of SWCNTs that can be capped by a $C_{60}$ molecule. It is important to note that the two different types of carbon nanotubes result essentially from the way the [60] fullerene molecule is bisected. If the fullerene molecule is bisected normal to a five-fold axis, the armchair configuration is formed. However, if the fullerene molecule is bisected normal to a three-fold axis, the zigzag configuration is formed.

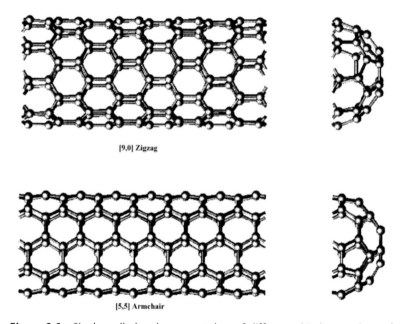

[9,0] Zigzag

[5,5] Armchair

**Figure 9.6** Single-walled carbon nanotubes of different chirality can be end capped by one half of a $C_{60}$ molecule. (Top) the zigzag (9,0) tube and (bottom) the armchair (5,5) tube.

Considering the SWCNTs as a 1-D fullerene, or a 1-D crystal, enables the definition of a unit cell that can be translated along the tube axis. For all tube types, such a translational unit cell is cylindrical in shape. The 2-D graphene sheet lattice model shown in Figure 9.7 is an excellent method to illustrate how the unit cell of a SWCNT can be constructed.

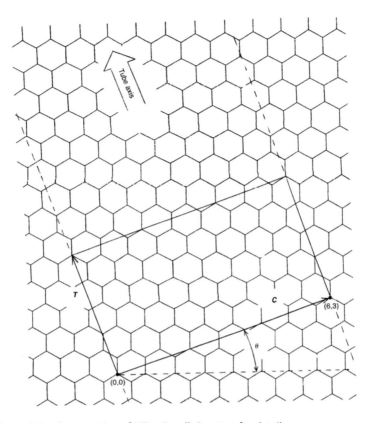

**Figure 9.7**   Construction of NT unit cell. See text for details.

As shown in the figure, the method involves drawing a straight line through the carbon atom selected as the origin (0,0), normal to the chiral vector ($C_h$) and extending this line until it intersects an exactly equivalent lattice point. The length of the tube transitional unit cell is the magnitude of the vector (**T**) as shown in Figure 9.7 for the case of a (6,3) nanotube. Figure 9.8 shows the transitional unit cell for (a) a (5,5) armchair tube, and (b) a (9,0) zigzag tube. From Figures 9.7 and 9.8, it is clear that the length of the translational unit cell depends not only on the diameter of the tube, but also on the tube chirality. As we have shown before, while both the (5,5) and the (9,0) tubes are equal in diameter (both can be capped by half of a $C_{60}$ molecule), the length of the unit cell is longer for the zigzag tube. While the length of the unit cell in the (5,5) armchair tube is equal to $a$ (2.461 Å according to Eq. 9.2), the unit cell length equals $a\sqrt{3}$ (4.263 Å) in the case of the (9,0) zigzag tube (see Figure 9.8).

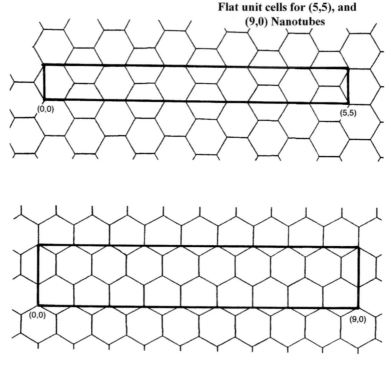

**Figure 9.8** Unit cells for a (5,5) armchair nanotube (upper) and a (9,0) zigzag nanotube (lower) shown on a flat graphene sheet.

Expressions have been derived for the length of the unit cell ($T$) in terms of the magnitude of the chiral vector ($C_h$) and the highest common divisor ($d_H$) of the coordinate multiples $n$ and $m$. It was shown that

$$T = \frac{C_h}{d_H}\sqrt{3} \text{ , if } n - m \neq 3qd_H \qquad (9.7)$$

while

$$T = \frac{C_h}{d_H}\frac{1}{\sqrt{3}} \text{ , if } n - m = 3qd_H \qquad (9.8)$$

It was also shown that for a tube specified by ($n,m$), the number of carbon atoms per unit cell of a tube is $2N$ such that

$$N = (n^2 + nm + m^2)/d_H \text{ if } n - m \neq 3qd_H \qquad (9.9)$$

and

$$N = (n^2 + nm + m^2)/3d_H \text{ if } n - m = 3qd_H \qquad (9.10)$$

where $q$ is, again, an integer.

These expressions enable the determination of the unit cell length and number of carbon atoms in there. For typical experimentally observed SWCNTs, the diameters range between 1 nm and 30 nm. This makes unit cells very large. For example, a 10 nm diameter tube would have a unit cell length in the range of 50~55 nm containing around 65,000 carbon atoms.

We have mentioned earlier that the smallest 0-D fullerene is the $C_{20}$ molecule. The question regarding the smallest 1-D fullerene or carbon nanotube has also been raised. The smallest carbon nanotube reported was a 0.4 nm in diameter confined inside an 18 shells carbon nanotube. Figure 9.9 shows the transmission electron micrograph and a superimposed atomic model of the nanotube.

**Figure 9.9** High resolution transmission electron micrograph and a superimposed atomic model of a 0.4 nm diameter nanotube confined in an 18 shell MWCNT. Reproduced with kind permission from Qin et al., *Nature*, 2000, 408, 50. Copyright Nature Publishing Group, 2000.

In year 2001, Stojkovic and co-workers describe how $sp^2$ carbon, in regular nanotubes can be replaced by $sp^3$ carbon to produce extremely small-diameter (in the range of 0.4 nm) carbon nanotubes with only minimal bond-angle distortion. These tubes were shown to have chiral multiples of either a zigzag (3,0) or an armchair (2,2), and were predicted to have a stiffness higher than that of traditional $sp^2$-bonded carbon nanotubes, therefore, forming the stiffest 1-D systems known. Figure 9.10 shows the calculated, relaxed structures of (a) the (3,0), and (b) the (2,2) $sp^3$ nanotubes both a single unit cell and a space-filling model of the tubular structures.

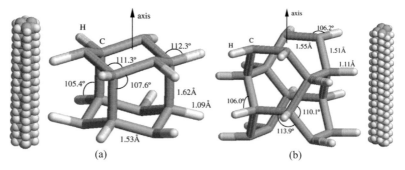

**Figure 9.10** The calculated, relaxed structures of (a) the (3,0), and (b) the (2,2) $sp^3$ nanotubes both a single unit cell and a space-filling model of the tubular structures. Reproduced with kind permission from Stojkovic et al., *Phys. Rev. Lett.*, 2001, 87, 125502. Copyright American Physical Society, 2001.

The symmetry of SWCNTs has also been thoroughly investigated and reported. While the symmetry of all carbon nanotubes is described by the space groups in 1-D space groups (line groups), SWCNTs *can still be classified* into two distinct groups from a symmetry viewpoint. One group, that includes armchair and zigzag nanotubes, can be represented by *symmorphic* space groups, meaning that none of the symmetry operations requires rotation and translation operations to be combined. This group is referred to as *achiral* tubes. The second group that includes chiral tubes, can be represented by *non-symmorphic* space groups in which some of the symmetry operations involves both rotation and translation operations. Figure 9.11 shows the symmetry elements for chiral, zigzag, and armchair carbon nanotubes. As shown in the figure, all carbon nanotubes possess rotation axes normal to their surface passing through the hexagon center and the midpoint of C–C bonds. In addition, only achiral (zigzag and armchair) nanotubes possess reflection, glide, and rotation–reflection symmetry elements.

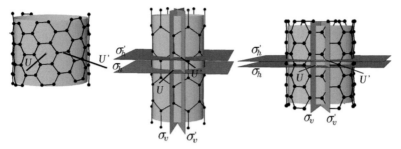

**Figure 9.11** Symmetry elements for chiral (8,6) nanotube (far left), zigzag (6,0) (middle), and armchair (6,6) (far right) nanotubes. U, and U' denote rotation axes, σ denotes mirror planes, σ'$_v$ is a glide plane, and σ'$_h$ is a rotation-reflection plane. Reproduced with kind permission from Damnjanovic et al., *Phys. Rev. B*, 1999, 60, 2728. Copyright American Physical Society, 1999.

Regarding the symmetry groups for achiral SWCNTs $(n,n)$ or $(n,0)$, it was shown that since all such tubes possess rotation symmetry axes in addition to mirror and/or glide planes (see Figure 9.11), they belong to either $D_{nh}$ or $D_{nd}$ groups. Assuming that all achiral tubes have an inversion center, it follows that achiral tubes with $n$ even belong to the $D_{nh}$ group, and that achiral tubes with $n$ odd belong to the $D_{nd}$ group. Hence, a (5,5) armchair or a (5,0) zigzag tube should belong to the $D_{5d}$ symmetry group. Also, a (6,6) armchair or a (6,0) zigzag tube belong to the $D_{6h}$ symmetry group.

Regarding the symmetry of chiral $(n,m)$ nanotubes the determination of the symmetry group is not as straight forward as in the case of achiral tubes. Here we will only outline the general method used to determine the symmetry group. Since chiral tubes do not possess any mirror planes, their symmetry should belong to the C symmetry groups. In addition, due to their nonsymmphoric symmetry, their basic symmetry operation $R_- = (\psi/\tau)$ involves a rotation by an angle $\psi$ followed by a translation $\tau$. This operation corresponds to the vector $\boldsymbol{R}$ such that

$$\vec{R} = p\vec{a}_1 + q\vec{a}_2 \tag{9.11}$$

Here, $(p,q)$ are the coordinates reached when the symmetry operation $(\psi|\tau)$ acts on an atom at the coordinates $(0,0)$. The values of $p$ and $q$ are given by,

$$mp - nq = d_H \qquad (9.12)$$

with the condition that

$$q < m/d_H \text{ and } p < n/d_H \qquad (9.13)$$

it can be shown that

$$\psi = 2\pi \frac{\Omega}{Nd_H} \text{ and } \tau = \frac{Td_H}{N} \qquad (9.14)$$

where $\Omega$ is defined as

$$\Omega = \frac{p9m + 2n) + q(n + 2m)}{(d_H/d_R)} \qquad (9.15)$$

The value of $d_R$ depends on the tube chirality. For tubes where $n-m$ is not a multiple of $3d_H$, $d_R = d_H$, whereas for tubes where $n-m$ is a multiple of $3d_H$, $d_R = 3d_H$. For chiral tubes with $d_H = 1$, the symmetry group belong to the $C_{N/\Omega}$ group. For chiral tubes with $d_H \neq 1$, the symmetry belongs to the group expressed by the direct product $C_{d_H} \otimes C'_{N/\Omega}$.

## 9.3 Multi-Walled Carbon Nanotubes

As we discussed earlier, the first carbon nanotubes to be discovered were multi-walled nanotubes (MWNTs). As shown in Figures 9.9 and 9.13, MWCNTs consist of multiple concentric single-walled nanotubes one inside the other. Typically, inner diameters of readily available MWNT range between 1 nm and 3 nm while outer diameters range between 2 nm and 20 nm. MWCNTs with larger outer diameters reaching hundreds of nanometers also exist. The important question regarding the structural correlation between the successive layers in a MWCNT was addressed and investigated. For MWCNTs, the interlayer distance between concentric tubes was measured using high resolution transmission electron microscopy and was reported to be around 3.4 Å. This means that the circumference of successive tubes should increase by about 21 Å. It can be easily seen that this is not possible for zigzag tubes that require the successive tubes circumference to increase by precise multiples of 0.246 nm that is the width of a single hexagon. For the zigzag MWCNTs, the closest approximation for the measured interlayer separation is when successive tubes circumferences differ in nine hexagons producing an inter-tube separation of 3.52 Å. In this case, the traditional ABAB stacking of perfect graphite can only be maintained over short distances in the circumferential direction. Figure 9.12 schematically illustrates a three-layer zigzag MWCNT. The figure also illustrates how interfacial dislocations can be introduced to accommodate hoop stresses.

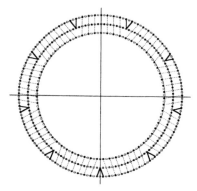

**Figure 9.12** Schematic illustration of a three-layer zigzag MWCNT as proposed by Zhang et al. The figure illustrates how interfacial dislocations (bold lines) can be introduced to accommodate hoop stresses. Reproduced with kind permission. Copyright Elsevier, 1993.

In the case of armchair tubes, however, the length of the repeating unit is 0.426 nm, and five times that would be very close to the required increase in the successive tube circumference (2.1 nm). Hence, multi-walled armchair structures can be assemblies in which the ABAB graphitic stacking sequence is maintained keeping the interlayer close to 3.4 Å. For chiral tubes, the restrictions on multi-walled tubes are more complicated and, generally speaking, it was concluded that it is not possible to have two consecutive layers having exactly the same chiral angle and separated by the exact interlayer spacing of 3.4 Å. End caps in MWCNTs are not as simple or well understood as those in SWCNTs. Figure 9.13 shows a typical MWCNT end cap.

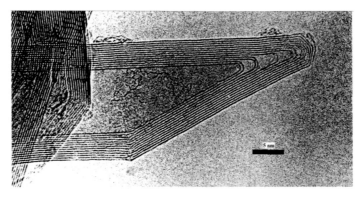

**Figure 9.13** High resolution transmission electron micrograph shows a typical multi-walled carbon nanotube end cap. Reproduced with kind permission. Copyright Royal Society of Chemistry, 1994.

## 9.4 Double-Walled Carbon Nanotubes

A special case of MWCNTs that dragged a lot of attention is the DWCNTs. While DWCNTs were among the first nanotube to be reported, attention was really paid to their properties later as the field of nanotechnology started to be better understood. With their mean outer diameter ranging between 9 Å and 18 Å and only two tubes, DWCNTs provide an excellent model system and an opportunity to investigate the physics of nanotubes on both experimental and theoretical levels. They are also expected to exhibit unique properties compared to those of SWCNTs and MWCNTs. Such unique properties include mechanical, thermal, electrical, and chemical properties leading to unique ability to store hydrogen for energy applications as well as unique ability to be functionalized with certain amino acids.

The fact that DWCNTs are dragging lots of attention due to their unique properties compared to these of SWCNTs and MWCNTs should, indeed, elucidate the core idea of nanosystems and nanotechnology as we discussed earlier. From the surface nature viewpoint, SWCNTs, DWCNTs, and MWCNTs should be essentially identical and similar to other forms of fullerenes. However, the physics of that particular cylindrical fullerene with only two layers, and outer diameter ranging between 0.9 and 1.8 nm (the double-walled nanotubes) is still unique. It must be that the uniqueness of DWCNTs is because the system scale is resonating with the important length scales defining nanosystems as we discussed before.

## 9.5    Production of Carbon Nanotubes

The methods of carbon nanotube production can mainly be classified into two groups. One group is based upon sublimation/ condensation of graphite and the other group is based upon pyrolysis of hydrocarbons. In both categories, the processes involved are catalytic in nature. In the sublimation-/condensation-based methods, the energy required for the sublimation process is provided by different sources such as arc discharge, laser radiation, resistive heating, or by concentrated solar radiation. The major disadvantages of all methods of this group include large energy consumption, relatively low yield of nanotubes, high cost due to the high cost of graphite, complications due to the required high vacuum and controlled atmosphere, and the difficulties of automation and up-scaling. Pyrolysis-based methods, on the other hand, are relatively much less complicated. This group of methods does not require high temperatures, can proceed under atmospheric pressure, can be run continuously, and provides a comparatively high yield. In addition, pyrolysis-based methods enable better control of the nanotube diameter and produce longer nanotubes than those produced by sublimation/condensation methods. The major disadvantage of pyrolysis methods is that they produce lower quality nanotubes having structural imperfections.

In the following sections, we will discuss different production methods belonging to the sublimation/condensation group as well as those belonging to the pyrolysis group.

### 9.5.1    The Arc-Discharge Method

Original production of carbon nanotubes was carried out in the same arc-discharge method used to produce spherical fullerenes where a DC arc is used to evaporate carbon electrodes in vacuum or inert atmosphere. This method is one of the sublimation/ condensation production methods. The only difference is that in nanotube production, the carbon electrodes were kept apart at a distance instead of being in continuous contact as in the case for fullerene production. The initial carbon nanotube yield using this method was rather poor. Subsequent studies showed that a number of factors are important in producing a high yield and good quality of carbon nanotubes. These factors included the pressure of the inert atmosphere inside the evaporation chamber, the stability of the plasma field, quality and composition of graphite electrodes, and efficient cooling of the electrodes and chamber. The most important factor of these is the pressure of helium inside the production chamber which appears to be optimized at 500 Torr. It is interesting to note that in this method if the helium pressure was kept under 100 Torr, the process mainly produces [60] fullerene molecules. This point elucidates the extremely important role played by the production atmosphere in determining which form of fullerene is most stable, and hence is the main product of the process. The fact that simulation studies (usually conducted in pure vacuum atmosphere) showed that spheroidal geometries of giant fullerenes are more favorable than tubular geometries might be clouded by the eliminated atmosphere factor.

Regarding the arc current, it is now known that maintaining the current as low as possible enable maintaining stable plasma and increases the nanotube production yield. In addition, a graphite rod containing metal catalyst (Ni, Fe, Co, Pt, Pd, etc.) used as the anode with a pure graphite cathode was found to increase the nanotube yield of the process.

The maximum yield is achieved with Ni-Co. It was also observed that the addition of sulfur increases the catalytic effect of other metal catalysts. Efficient cooling has also been shown to enable avoiding excessive sintering of the soot—an essential condition for good quality nanotubes. High yield of SWCNTs was also achieved in the arc-discharge method by using a bimetallic Ni-Y catalyst in helium atmosphere. The efficiency by this method was sharply increased by positioning the two graphite electrodes at 30° angle instead of the conventional 180° alignment. This new arrangement is referred to as the *arc-plasma-jet method* and it typically yields SWCNTs at a rate of 1 g/min.

The morphology of the produced nanotube was also found to depend on the nature of the chamber atmosphere. MWCNTs were primarily formed in methane atmosphere. The important role of hydrogen presence in the formation and production of MWCNTs was confirmed by multiple studies. It is also interesting to note that spherical fullerenes will not form in an atmosphere of a gas including hydrogen atoms. Water as an atmosphere for the arc-discharge method was also used to produce well graphitized multi-wall carbon nanotubes. In addition, a new morphology of carbon nanostructures shaped in the form of a petal was found to form in such atmosphere. The production of DWCNTs by arc-discharge method in hydrogen atmosphere under similar operating conditions was also reported. Production of high-quality double- and single-walled nanotubes was achieved by a *high temperature–pulsed arc-discharge* method which utilizes a dc-pulsed arc discharge located inside a furnace to maintain homogeneous production condition.

Carbon nanotubes grown by the arch discharge method are, in general, not aligned but are rather produced in the form of a web. Partially aligned SWNTs can be grown by convection or a directed arc plasma method. Because of the high production temperature of the arc-discharge method, the produced carbon nanotubes are expected to be of high quality.

## 9.5.2   Other Condensation Methods

Carbon nanotubes could also be produced through the condensation of carbon vapor in the absence of an electric field. The process starts with the evaporation of a graphite target and then the carbon vapor would condense into carbon nanotubes, mainly single-walled, on a cooled substrate under the appropriate conditions. Earliest reports involving such a method for carbon nanotubes production came from a group at the Russian Academy of Science who used an electron beam to evaporate graphite in high vacuum ($10^{-6}$ Torr). The carbon vapor condensed on a quartz substrate producing, among other forms, what was later described as imperfect MWCNTs. The condensation of carbon vapor was also used to produce carbon nanotubes after evaporating carbon by resistively heating carbon foils and condensing the vapor on freshly cleaved highly oriented pyrolytic graphite (HOPG) substrate under high vacuum ($10^{-8}$ Torr). The produced nanotubes were essentially SWCNTs with diameters ranging between 1 nm and 7 nm laser beams (typically a YAG or a $CO_2$ laser) were also used to evaporate carbon in an inert gas atmosphere to produce carbon nanotubes in the same method as well. Concentrated sunlight was also reported as an applicable graphite sublimation method to produce carbon nanotubes.

### 9.5.3 The HiPco Process and Other Pyrolytic Methods

For over a century, it has been known that filament-like carbon species can be produced by catalytic decomposition of carbon-containing gas on a hot surface. This phenomenon was first discovered by the Schultzenbergers in 1890 while experimenting with the passage of cyanogens over red-hot porcelain. The phenomenon was re-investigated in the 1950s and it was established that carbonous filaments can be produced by the interaction of a wide range of hydrocarbon and other carbon containing (such as CO) gases with metal surfaces such as iron, nickel, platinum, and cobalt. In case of carbon monoxide source, the technique mainly depends on the CO disproportionation reaction that can be described as follows:

$$2CO_{(g)} \Leftrightarrow C_{(s)} + CO_{2(g)} \qquad (9.16)$$

The very early, and unrealized, production of carbon nanotubes by Endo and co-workers from benzene vapor was based on this process. In 1999, the optimization details of a large-scale production method of quasi-aligned carbon nanotube bundles was reported. The method includes catalytic decomposition of acetylene over well-dispersed metal particles embedded in commercially available zeolite at a lower temperature (in the range of 700 °C). The produced tubes were mainly homogeneous, perfectly graphitized multi-walled tubes with inner diameter ranging from 2.5 nm to 4 nm, and an outer diameter ranging from 10 nm to 12 nm.

In 2001 Smalley's group at Rice University rediscovered the technique and developed a new method which they called "HiPco" that utilized high pressure carbon monoxide as a precursor for

SWCNT mass production. In this process SWCNTs are grown in a high-pressure (in the range of 30–50 atm), high-temperature (in the range of 900–1100 °C), flowing CO atmosphere. Iron is added to the gas flow in the form of iron pentacarbonyl [$Fe(CO)_5$]. Upon heating, the $Fe(CO)_5$ decomposes and the iron atoms condense into clusters. These clusters serve as catalytic particles upon which SWNT nucleate and grow in the gas phase via CO disproportionation reaction. The technique produces highly pure SWCNTs at rates of up to 450 mg/h.

Since then, many other pyrolytic methods have been described in which carbon nanotubes can be produced. For example, modified, self-regulated arc-discharge method was also used to decompose liquid hydrocarbons in carbon nanotube structures. An electric furnace was also used to commercially produce carbon nanotubes from toluene. More recently, production methods based upon decomposition of ethanol (with ferrocene as a catalyst) in a vertical furnace, and over silica coated cobalt catalyst were reported. Acetylene diluted with argon was also reported to be a good precursor for pure SWNT production.

## 9.6 Purification of Carbon Nanotube

Purification of carbon nanotubes is, typically, based upon oxidative treatment using oxygen or other oxidants such as nitric acid or hydrogen peroxide to remove carbonaceous impurities as well as metal catalyst impurities. In order to reduce carbon nanotube damage due to oxidative purification treatments, other acidic treatments were suggested using hydrochloric acid (HCl). Such alternative acidic treatments while shown to be less damaging to the nanotubes, they are also considered less effective in removing the impurities. More recently, a room temperature, liquid bromine based treatment was reported to be more effective in removing impurities and less damaging to the nanotubes than other treatments. Table 9.1 shows different carbon nanotube purification treatments. Figure 9.14 shows an electron micrograph of carbon nanotubes before and after purification.

**Table 9.1**    CNTs Purification methods used in the literature

| Purification procedure | % by weight iron remaining after purification (lit., %) | % by weight iron remaining after purification[a] (this work, %) | NIR luminescence intensity[b] (this work) |
|---|---|---|---|
| HCl (35%), 4 h at 60°C | <1 | 10.2–14.4 | Good |
| Microwave radiation for 2 min, then HCl (35%), 4 h at 60°C | 9 | 10.8–12.6 | Good |
| Microwave radiation for 20 min, then HCl (35%), 4 h at 60°C | 7 | 10.5–12.5 | Good |
| $H_2SO_4$ (98%) + $HNO_3$ (70%), 4 h at 60°C | N/A | 9.7–14 | Poor |
| $H_2SO_4$ (25%), 10 min at 20°C | N/A | 13.9–14.8 | Good |
| $HNO_3$ (10%), 4 h at 60°C | <1 | 0.6–0.8 | Poor |
| $O_2$ +$SF_6$(g) at 200–400°C for 3–7 days, then HCl (35%), 12 h at 60°C | 1.5 | 3.0 | Good |
| $Br_2$ (l), RT, 4 h | N/A | 2.8–3.6[d] | Good |
|  |  | 1.6–1.8[e] | Poor |

[a]Iron analysis by ICP-AE.
[b]Comparison with the NIR Luminescence intensity for the original unpurified, raw SWCNTs (100%); an intensity greater than 25% of the original is considered as "good," below 5% as "poor," excitation laser at 660 nm with luminescence observed in the 900–1600 nm range.
[c]$H_2SO_4$ mixed with $H_2O_2$ (30%) at 0°C then diluted with water.
[d]After a first purification cycle.
[e]After a second purification cycle.

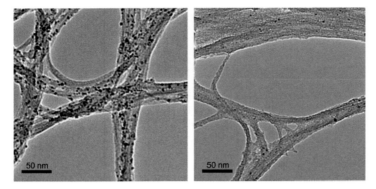

**Figure 9.14** Transmission electron micrograph of as-produced SWCNTs (left) and bromine purified SWCNTs purified to a residual iron content of ca. 3% by weight (right). Reproduced with kind permission from Mackeyev et al., *Carbon*, 2007, 45, 1013. Copyright Elsevier, 2007.

## 9.7 Mechanical Properties of 1-D Fullerenes

Mechanical properties of linear 1-D fullerenes (SWCNTs and MWCNTs) were investigated theoretically in order to determine their mechanical characteristics or constants such as elastic modulus and ultimate tensile strength. Table 9.2 lists the values for SWCNTs, MWCNTs, and structural steel for comparison. It is important to note that experimental verification of such theoretical values was never done due to the impossibility of measuring stress–strain curves for individual tubes. However, elastic modulus of individual tubes determined by non-traditional methods such as bending resonance under electron beams yielded values very close to the theoretical values.

**Table 9.2** Theoretical mechanical properties of SWCNTs and MWCNTs

| Materials | Axial elastic modulus | Axial tensile strength |
|---|---|---|
| SWCNT | 1.25 TPa | 150 GPa |
| MWCNT | 1.2 TPa | 150 GPa |
| Structural Steel | 210 GPa | 0.4 GPa |

## 9.8    Raman Scattering of Single-Walled Carbon Nanotubes

The Raman spectrum of SWCNTs was also investigated both theoretically and experimentally. Lattice dynamics and phonon symmetry investigations for SWCNTs show that there are 8 Raman active modes for achiral tubes and 14 Raman active modes for chiral tubes. The symmetry species of the Raman active modes can be classified as follows:

Zigzag tubes:        $\Gamma = 2A_{1g} + 3E_{1g} + 3E_{2g}$        (9.17)

Armchair tubes:    $\Gamma = 2A_{1g} + 2E_{1g} + 4E_{2g}$        (9.18)

Chiral tubes:        $\Gamma = 3A_1 + 5E_1 + 6E_2$        (9.19)

Figure 9.15 shows the displacement vector for Raman active modes of (a) armchair and (b) zigzag SWCNTs. As shown in the figure, for armchair tubes, the two totally symmetric ($A_{1g}$) modes originate from the radial breathing mode and the displacement of atoms in the circumferential (transverse) directions. For zigzag tubes, the two totally symmetric ($A_{1g}$) modes originate from the radial breathing mode and the longitudinal (shear) displacement of the carbon atoms. In chiral tubes, however, both longitudinal and transverse phonons are fully symmetric ($A_1$).

In general, the Raman spectrum of SWCNTs shows two classes of peaks: a low wavenumber class (usually less than 300 cm$^{-1}$) resulting from radial displacements of the carbon atoms (radial breathing mode of the tube) and a high wavenumber class resulting from tangential displacements of the carbon atoms (tangential or surface modes) in the range 1350 to 1580 cm$^{-1}$. Second order modes can also be observed in the Raman spectrum around 2700 cm$^{-1}$. Figure 9.16 shows a typical spectrum of SWNT showing the RBM, the tangential modes, and the second order modes as measured using a near infrared 785 nm wavelength laser excitation and recorded at ambient conditions. The higher wavenumber modes include the D-band (around 1360 cm$^{-1}$) which is related to structural defects in the nanotube, the G-band (three modes belonging to the A, E$_1$, and E$_2$ symmetry in the range from 1550 cm$^{-1}$ to 1605 cm$^{-1}$) and the G′ band (a combination mode around 2700 cm$^{-1}$). In the figure, the features marked with '*' at 303, 521 and 963 cm$^{-1}$ are from the Si/SiO$_2$ substrate.

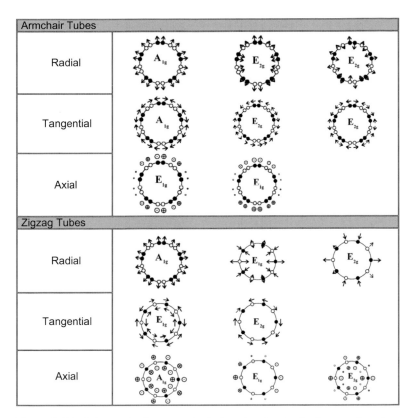

**Figure 9.15** Schematic diagrams depicting the atomic displacement vectors corresponding to all Raman active vibration modes in armchair (*n,n*), and zigzag (*n,0*) single-walled carbon nanotubes.

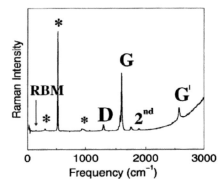

**Figure 9.16** Full Raman spectrum of single-walled carbon nanotube taken with excitation laser (785 nm). Reproduced with kind permission from Jorio et al., *Phys. Rev. Lett.*, 2001, 86, 1118. Copyright American Physical Society, 2001.

A number of studies were devoted to investigate the correlation between Raman spectrum characteristics and SWCNTs structure and properties. The position of the radial breathing mode (v) in SWNT bundles was reported to correlate linearly with the tube diameter ($d$) according to the relationship

$$v(\text{cm}^{-1}) = \frac{223.75}{d(\text{nm})} + 14 \qquad (9.20)$$

When isolated individual SWNTs were investigated, the correlation between the RBM position and tube diameter was reported to be

$$v(\text{cm}^{-1}) = \frac{248}{d(\text{nm})} \qquad (9.21)$$

And later reported to be

$$v(\text{cm}^{-1}) = \frac{214.4 \pm 2}{d(\text{nm})} + 18.7 \pm 2 \qquad (9.22)$$

The difference in the Raman position of the RBM was related to tube/tube interaction within a tube bundle. Tube interaction with their supporting substrate or their environment would also cause a shift in the tube radial breathing modes due to perturbation effects as we discussed earlier. Such perturbation effects can cause significant shifts in the radial breathing mode of SWCNTs rendering the aforementioned equations useless.

Intensities of radial breathing modes were also found to depend in the energy of the excitation laser due to Raman resonance effects. Metallic tubes with $|n - m| = 3q$ and semiconducting tubes with $|n - m| = 3q \pm 1$ (where $q$ is an integer) would resonate at different excitation laser energies according to their diameter which controls their electronic structure. It was established that once the excitation laser energy is within 0.1 eV of the van Hove singularity[1] inter-band transition ($E_{ii}$) for a particular tube, the Radial breathing mode of that particular tube with exhibit a resonance Raman effect and its RBM band will be prominent in the Raman spectrum. Hence certain

---

[1]For readers without physics background, an elementary solid-state physics reference can be consulted for further information on the van Hove singularities. However, the point here is to realize that depending o the excitation laser wavelength certain types (metallic versus semiconducting) with certain diameters nanotubes will be in resonance, hence their RBM will be more prominent in the spectrum.

nanotubes can be more observable using the typical argon ion green laser (wavelength 514.5 nm) while other nanotubes can be more observable using a typical solid-state laser (wavelength 780 nm).

Raman dispersion effects was clearly observed in the D-band and its second order G'-band features of both metallic and semiconducting SWCNT Raman spectra. Due to dispersion effects, the observed frequency of the D-band ranges between 1250 cm$^{-1}$ and 1450 cm$^{-1}$. The frequency of the second order feature G', however, was reported to range between 2500 cm$^{-1}$ and 2900 cm$^{-1}$ also due to dispersion effects. The observed dispersion effect in the frequency of both bands showed a linear dependence on the laser excitation energy ($E_L$) within the range $1.0 < E_L < 4.5$ eV with a plateau or step-like feature around excitation energy of 2 eV. The step-like feature was related to electron–phonon coupling effects. The frequency of the G' ($\omega_{G'}$, in cm$^{-1}$) linear dependence on the excitation laser energy ($E_L$ in eV) could be fitted to the equation:

$$\omega_{G'} = 2420 + 106E_L \qquad (9.23)$$

Figure 9.17 shows the experimental results measured for the dispersion effect on the peak position of the G' band for laser excitation energies ranging between 1.5 and 3.0 eV.

Recently, dispersion effect in graphitic material in general, and in carbon nanotubes in particular, was related to a double-resonance scattering effect. Figure 9.18 shows the experimental results versus calculations—based upon double-resonance theory—for the dependence of the D-band on laser excitation energy. It is clear from the figure that the calculations are in good agreement with experimental results reported by several groups.

Another important capability of the Raman technique is its ability to discriminate metallic and semiconducting nanotubes based on the shape of the G-band. The crucial point to note here is the fact that in a nanosystem such as a SWCNT, slight changes in the system features, chirality for example, leads to measurable changes in the system's properties that can be investigated using the Raman technique. It is important to note that while Raman investigation of SWCNTs started as early as their discovery in the early 1990s, the richness of the technique capabilities as a diagnostic tool for carbon nanotubes is still being explored.

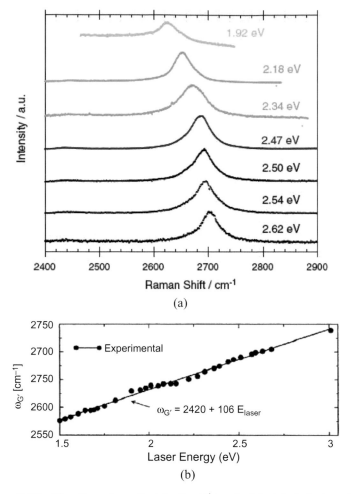

**Figure 9.17** The dispersion effect in the G′-band of isolated single-walled carbon nanotubes. (a) Raman spectra; reproduced with kind permission from Shimada et al., Ref. [106]. Copyright Elsevier, 2005. (b) G′-band position plotted as a function of excitation laser energy; reproduced with kind permission from Filho et al., *Phys. Rev. B*, 2001, 63, 241404. Copyright American Physical Society, 2001.

Polarized Raman also plays a major role in investigating the structure of carbon nanotube-based systems. The technique provides invaluable information regarding the orientation of nanotubes. The intensity of Raman active mode depends on the polarization scattering settings as measured in reference to the tube axial direction. Figure 9.19 shows polarized Raman spectra (collected in a backscattered ∥/∥ geometry) of an ultra-thin film (less than 60 nm) of oriented SWCNTs. It is clear from the figure that the intensity of the Raman active modes is highest when the laser polarization direction is parallel to the tube axis direction ($\theta = 0°$). The Raman active modes, however, disappear when the laser polarization direction is normal to the tube axis ($\theta = 90°$). Figure 9.20 shows the experimentally measured intensities (collected in a backscattered ∥/∥ geometry, a VV geometry) of all Raman active bands in an isolated SWCNT as a function orientation angle ($\theta$) between the excitation laser polarization and the tube axial direction.

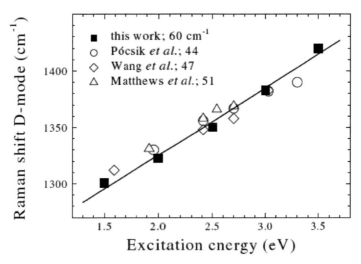

**Figure 9.18** Measured and calculated frequencies of the *D* band as a function of the excitation energy. The open symbols correspond to experimental data and the closed squares to the calculated phonon energies in double resonance. The line is a linear fit to the theoretical values with a slope of 60 cm⁻¹/eV, the numbers give the corresponding slopes for the data reported by each research group. Reproduced with kind permission from Thomsen and Reich, *Phys. Rev. Lett.*, 2000, 85, 5214. Copyright American Physical Society, 2000.

The ability of the Raman technique to measure orientation of nanotubes is essential to investigating nanostructured mesosystems. Figure 9.21 shows color coded polarized Raman maps depicting the orientation of SWCNTs within two different films. The map to the left (Figure 9.21a) is for a very well aligned film, while the map to the right (Figure 9.21b) shows that the film contains domains of well-aligned tubes separated by boundaries of misaligned tubes. Based on our previous discussions, it can be predicted that the electrical, thermal, optical, and most probably chemical properties of these two films will be different. However, this particular area is still a frontier to explore by systematic scientific investigation driven by very promising evidences.

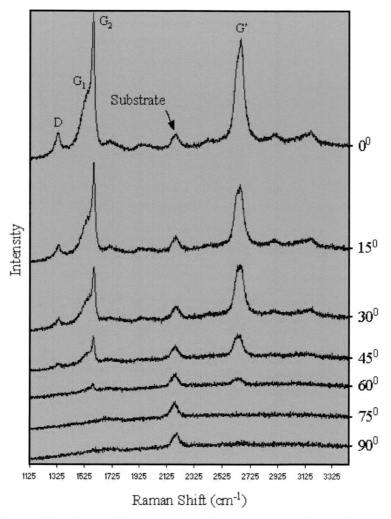

**Figure 9.19** Polarized Raman spectra collected in a backscattered ‖/‖ setup of an ultra-thin film ($t < 60$ nm) of oriented single-walled carbon nanotubes deposited on a substrate. Each spectrum is collected at different angle ($\theta$) between the polarization direction and the tube axis direction. Figure courtesy of Prof. M. S. Amer.

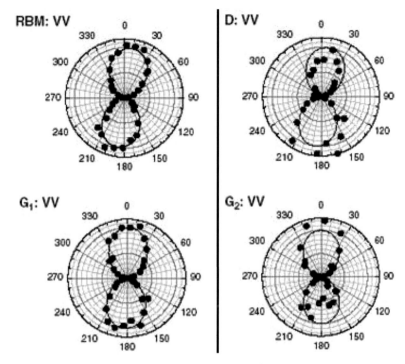

**Figure 9.20** Experimentally measured intensities (collected in a backscattered ∥/∥ geometry, a VV geometry) of all Raman active bands in an isolated single-walled carbon nanotube as a function orientation angle (θ). Reprinted with kind permission from Duesberg et al., *Phys. Rev. Lett.*, 2000, 85, 5436. Copyright American Physical Society, 2000.

## 9.9   Raman Scattering of Double- and Multi-Walled Carbon Nanotubes

The Raman spectrum of MWCNTs typically shows the D, G, and G′ bands. Radial breathing modes typical of SWCNTs are not part of a multi-walled tube. The simplest explanation for this can be given along the notion that radial breathing mode requires all carbon atoms to translate in-phase in the radial direction. The possibility for such a move to occur in multi-concentric tubes is very low.

DWCNTs on the other hand, provide the simplest system to study the perturbation effects on nanotubes especially the confinement due to interaction between concentric nanocylinders. Since SWCNTs can be either metallic (M) or semiconducting (S),

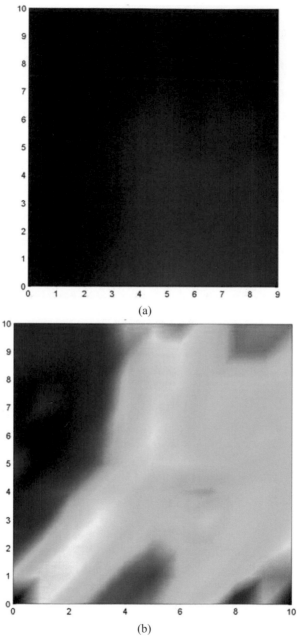

(a)

(b)

**Figure 9.21** Polarized Raman maps of single-walled carbon nanotube films on silicon substrates. (a) A 10 × 10 µm orientation map of a well aligned homogeneous film, and (b) a 10 × 10 µm map showing multi-domains with boundaries of unaligned regions. Figure courtesy of Prof. M. S. Amer.

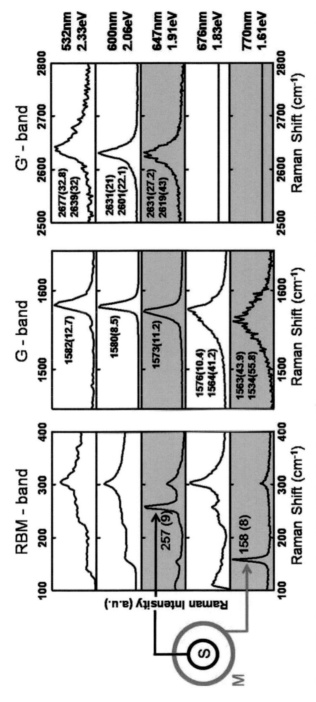

**Figure 9.22** Raman spectra of the RBM, G band, and G′ band features taken from an isolated S@M DWNT. By using two laser energies (shaded regions), we can detect the inner semiconducting ($E_L$ = 1.91 eV) and the outer metallic ($E_L$ = 1.61 eV) constituents of the same DWNT. The peak frequencies are marked followed by the corresponding FWHM in parenthesis. A blank frame means that no G′ data was acquired for 1.61 and 1.83 eV due to luminescence from the Si substrate. Reproduced with kind permission from Villalpando-Paez et al., *Nano Lett.*, 2008, 8, 3879. Copyright American Chemical Society, 2008.

DWCNTs can assume any of the four possible configurations (i.e., M@M, M@S, S@S, S@M). Each of these configurations would have different electronic properties, and hence, a different Raman spectrum. The configurations available for MWCNTs are still more versatile and sophisticated. Several investigations have utilized Raman scattering in order to understand the dependence of the electronic and optical properties of DWCNTs and MWCNTs on their configuration. Considering the recently developed techniques to remove end caps and shorten nanotubes, or in other words, tailoring nanotubes, and to selectively synthesize different types of single- and multi-layered nanotubes, such investigations would definitely have important implications in the fabrication of electronic devices using different types of semiconducting and metallic nanotubular interconnects. Figure 9.22 shows Raman spectra of the RBM, G band, and G′ band features taken from the same isolated DWCNT consisting of semiconducting tube in a metallic one. By using two laser energies (shaded regions), the inner semiconducting ($E_L$ = 1.91 eV) and the outer metallic ($E_L$ = 1.61 eV) constituents of the DWNT can be detected. The peak frequencies are marked followed by the corresponding full width half maximum (FWHM) in parenthesis. A blank frame in the figure means that no G′ data was acquired due to luminescence from the silicon substrate. It is also worth noting that once the structure of the DWNT was reversed, i.e., a metallic tube inside a semiconducting tube, the Raman spectrum of the tube was different. In this case both metallic and semiconducting tubes were excited—meaning that their radial breathing modes were observable— at the same excitation laser energy. Figure 9.23 shows the Raman spectrum of an isolated M@S DWNT (using laser energy = 2.33 eV) depicting the RBM, G, and G′ bands of both tubes.

Inter-tube interaction and its effect on tube properties (especially electrical, thermal, and optical) as reflected in the tube Raman activity and band shape and position is also illustrated by the observed shifts in the G′-band positions of SWCNTs and MWCNTs once they are mixed together. It was shown that once SWCNTs are mixed with multi-walled nanotubes, the position of the G′-band of both types of tubes exhibits a blue shift. The shift is linear as a function of SWNT weight (%) in the mixture. The mixtures were prepared by dissolving both types of tubes in dimethylformamide (DMF), mixing the solutions and sonicating the mixed solution for 30 minutes. Raman spectra were recorded at ambient conditions using a 514.5 nm argon ion laser after the mixtures were allowed to completely dry on glass slides. Figure 9.24 shows the blue shift exhibited by both SWCNTs and MWCNTs once allowed to mix together.

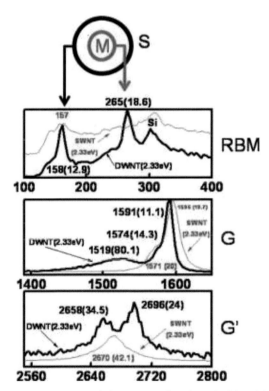

**Figure 9.23** Raman spectra of the RBM, G band, and G′ band features taken (using $E_L$ = 2.33 eV) from an isolated M@S DWNT. The RBM region shows that the inner and the outer tubes are simultaneously in resonance with the same laser line. Reproduced with kind permission from Villalpando-Paez et al., *Nano Lett.*, 2008, 8, 3879. Copyright American Chemical Society, 2008.

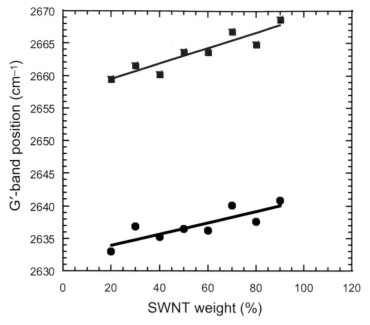

**Figure 9.24** Measured G′-band peak position for SWNT and MWNT mixtures as a function of SWNT weight% in the mixture. Figure courtesy of Prof. M. S. Amer, 2008.

## 9.10 Thermal Conductivity

Right after the observation of SWCNTs molecular dynamic calculations were conducted to calculate their thermal conductivity. The calculations performed on a (10,10) armchair isolated nanotube showed a very high thermal conductivity of 6600 W/m·K along the tube axial direction. Compared to the most highly conductive metals such as silver, gold, and copper with thermal conductivities in the range of 300 to 420 W/m·K, the thermal conductivity of isolated SWCNT is extremely high. Further studies, showed that thermal conductivity of SWCNT depends on the tube chirality, diameter, and length. The thermal conductivity drops as the tube diameter and/or length increases. In addition, MWCNT were found much less thermally conductors than SWCNT. Once bundled together, the thermal conductivity drops to the level of thermally insulating

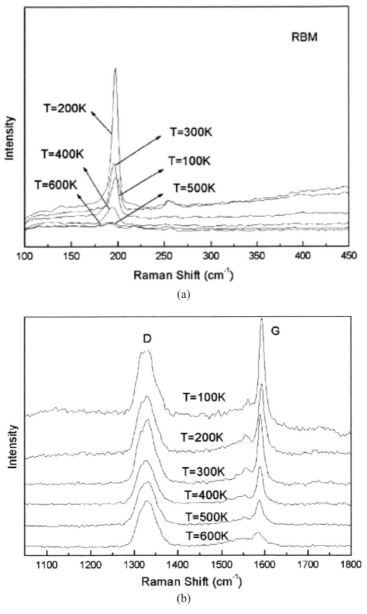

**Figure 9.25** The variation of the relative intensity of (a) RBM, and (b) D- and G-bands with changing temperatures for individual single-walled carbon nanotubes. Reproduced with kind permission from Zhou et al., *J. Phys. Chem. B*, 2005, 110, 1206. Copyright American Chemical Society, 2005.

materials. For example, a bundle of MWCNT thermal conductivity is 0.1 W/m·K. Unlike metals, where thermal conductivity is dominated by the flow of free (delocalized) electrons, the thermal conductivity of isolated SWCNT was found to be dominated by the phonon mechanism where heat is transported through the collective vibration of atoms in the system. Even in SWCNT with metallic nature, it was the phonon heat transfer mechanism that is dominating their high thermal conductivity. In fact, the predomination of the phonon mechanism explains the peculiar observation that bundles of MWCNTs are thermally insulating while an isolated SWCNT exhibit extremely high thermal conductivity.

Since the thermal conductivity of CNTs was attributed mainly to the phonon transport mechanism, investigating thermal properties of carbon nanotubes and the temperature dependence of their Raman active modes was the most promising experimental investigating technique. In general, all investigations have consistently reported two observations; first, that the intensities of all Raman active modes decrease with increasing the measurement temperature (see Figure 9.25), secondly, that the frequencies of all Raman active modes exhibit a linear red shift (softening) as the temperature is increased (see Figure 9.26). It was also reported that while the slope of the linear frequency dependence on temperature ($d\omega/dT$) is independent of the tube diameter for the G-band, it increases nonlinearly as the tube diameter increases for the radial breathing mode (see Figure 9.27).

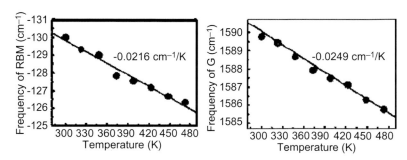

**Figure 9.26** Temperature dependence of the RBM (left) and G-band (right) of a suspended individual single-walled carbon nanotube. Reproduced with kind permission from Zhang et al., *J. Phys. Chem. C*, 2007, 111, 14031. Copyright American Chemical Society, 2007.

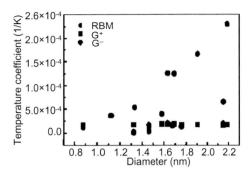

**Figure 9.27** Slope of frequency dependence on temperature (temperature coefficient $d\omega/dT$) as a function of tube diameter for RBM (solid circles), and the high and low frequency G-bands (squares and triangles, respectively) in single-walled carbon nanotubes. Reproduced with kind permission from Zhang et al., *J. Phys. Chem. C*, 2007, 111, 14031. Copyright American Chemical Society, 2007.

Knowing that the Stokes/anti-Stokes intensity ratio ($I_S/I_{AS}$) is related to the local temperature of a materials system. It could be tempting to assume that determining the local temperature in nanotubes is a straightforward process. The fact that radial breathing modes are typically located at wavenumbers lower than 300 cm$^{-1}$ making their experimental intensity ratio measurements possible and accurate increases such temptation. However, the resonance dependence of the radial breathing mode of SWCNTs on excitation laser energy, discussed earlier, put serious restrictions on such scientific temptation. The Stokes and anti-Stokes intensities of the radial breathing mode of in solution individual as well as in air bundles of SWCNTs were investigated. The intensities were measured using different excitation laser energies covering the range from 1.9 to 2.4 eV. They show an interesting dependence of both intensities on laser energy. Figure 9.28 shows Stokes (solid symbols) and anti-Stokes (open symbols) experimental resonance windows obtained for the same ($n,m$) SWCNT (RBM 244.4 cm$^{-1}$) dispersed in aqueous solution and wrapped with SDS (left) and in a bundle (right). It is clear from the figure that the Stokes/anti-Stokes intensity ratio does, indeed, depend on the excitation laser wavelength. A comparison between the right and left graphs also indicate that perturbation effects from the tube environment, also, have an effect on the measured ratio. This clearly indicates that due to resonance and perturbation effects, the Stokes/anti-Stokes ratio may not be characteristic of the tube temperature. While suggestions to utilize the calibrated linear dependence of phonons frequencies on temperature as a method to obtain sample temperature have been given, it is highly recommended to avoid the radial breathing modes due to their thermal coefficient dependence on exact tube diameter.

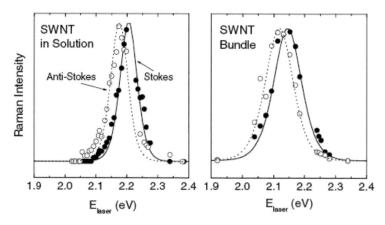

**Figure 9.28** Stokes (solid symbols) and anti-Stokes (open symbols) experimental resonance windows obtained for the same (*n,m*) single-walled carbon nanotube (RBM 244.4 cm$^{-1}$) dispersed in aqueous solution and wrapped with SDS (left) and in a bundle (right). Reproduced with kind permission from Fantini et al., *Phys. Rev. Lett.*, 2004, 93, 147406. Copyright American Physical Society, 2004.

## 9.11 Solvent Interactions

### 9.11.1 Solubility

In order to realize the superior properties of individual isolated CNTs, it was necessary to disperse them in solvents. This process required overcoming the very large depletion forces that are associated with linear fullerenes as we discussed in Chapter 5. The challenge is largest in case of SWCNTs due to their high aspect ratio. In addition, studies utilizing density functional theory (DFT) calculations investigating the electron density distribution in SWCNTs (see Figure 9.29) showed that the tubes lack the weak acidic hydrogens, hence, they are completely insoluble in virtually any solvent. This is why unlike spherical fullerenes, there is no data on the solubility limits of 1-D fullerenes. However, certain solvents were identified as at best capable of dispersing SWCNTs. Such solvents include *N,N*-dimethylformamide (DMF), 1,2-dichloroethane (DCE), dichlorobenzene (DCB), and *N*-methylpyrrolidone (NMP). Other solvents such as acetone and alcohols such as methanol, ethanol, and propanol were also reported to disperse SWCNTs effectively. SWCNTs could also be dispersed in aqueous solvents by wrapping the individual tubes with surfactant molecules (basically soap) such as sodium dodecyl sulfate (SDS) (see Figure 9.30).

**Figure 9.29** Electron density distribution for SWCNT as calculated using DFT method.

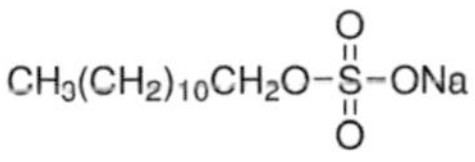

$$CH_3(CH_2)_{10}CH_2O-\overset{\displaystyle O}{\underset{\displaystyle O}{\overset{\|}{\underset{\|}{S}}}}-ONa$$

**Figure 9.30** The chemical structure of sodium dodecyl sulfate (SDS).

## 9.11.2 Effect on Solvents

The interaction of 1-D fullerene molecules with different liquids (or solvents) leading to structuring of the solvent was also confirmed by molecular dynamic simulation results. In order to investigate the effect of CNTs on the molecular structure of their environments, molecular dynamic simulations were utilized to calculate the equilibrium molecular structure of three different types of environments, namely, a non-interacting environment represented by an inert gas, a non-associated, non-polar fluid represented by carbon tetrachloride ($CCl_4$), and an associated, polar fluid represented by water ($H_2O$). Figure 9.31 shows the equilibrated structures of molecular dynamic simulation results for the three aforementioned cases. It is clear from the figure that the presence of the tubular fullerene (nanotube) does, indeed, have an effect on the molecular structure of the surrounding fluid. Layered molecular rings of the surrounding fluid can be clearly detected around the nanotube. The effect of the nature (or in other words the intra-molecular interactions within the fluid) is clearly demonstrated in the fact that the long-range interaction distance is different from one environment to the other. Equilibrated structures shown in Figure 9.31 also indicate that the long-range interaction effect of the nanotube on its environment decreases as the association among the environment molecules increases. Such results are perfectly in agreement with expectation since associative interactions among the fluid molecules would definitely resist any external influence resulting from the presence of the nanotube.

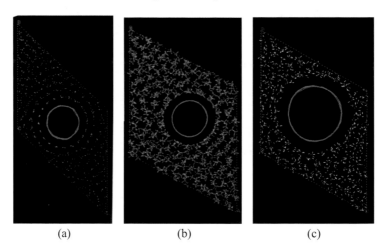

(a)                    (b)                    (c)

**Figure 9.31** Equilibrated structures of molecular dynamic simulation results for a single-walled carbon nanotube inserted in three different environments; (a) a Lennard Jones fluid (Ar, Kr), (b) a non-polar fluid ($CCl_4$), and a polar fluid ($H_2O$). Figure courtesy of Dr. J. Elliott, University of Cambridge, England.

## 9.12   2-D Materials Based on Single-Walled Carbon Nanotubes

Recently, 2-D films that are made of highly aligned SWCNTs were produced. Figure 9.32 shows the LB isotherm of the film indicating the monolayer nature of the film. Figures 9.33 and 9.34 show scanning tunneling microscopy (STM) images of a 1.67 × 1.67 nm and a 25 × 25 nm areas of the monolayer film, respectively. The high alignment of the individual tubes is very clear from the STM images. The compressive stress strain curve of the film is shown in Figure 9.35. It is important to note that the film could be compressed only normal to the tube axial direction due to experimental setup. As shown in Figure 9.35, the film exhibits a transverse stiffness (transverse to the tubes axial direction) of 10.21 MPa and it deviates from linear elasticity at applied strain close to 33% showing a minor strain stiffening behavior. Comparing the mechanical behavior of 2-D films based on 0-D and 1-D fullerenes (compare Figures 9.35, 8.42, and 8.43) it is clear that 2-D films based on 0-D fullerene ($C_{60}$) are much stiffer but lacks the strain stiffening behavior. These films also show a linear elastic behavior under compression up to 20% strain followed by film collapse unlike SWCNTs based films that show a higher linear elasticity (up to 33%) followed by strain stiffening behavior before collapse close to 50% strain.

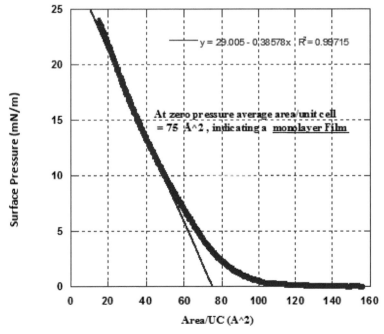

**Figure 9.32**  LB isotherm for the produced 2-D films based on highly aligned SWCNTs.

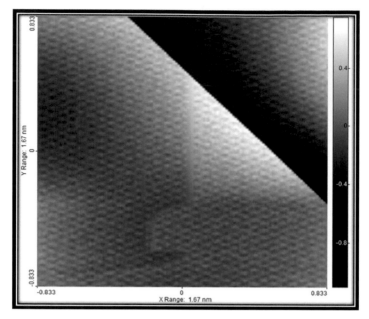

**Figure 9.33** 1.67 × 1.67 nm STM image showing the individual SWCNTs in the monolayer film produced.

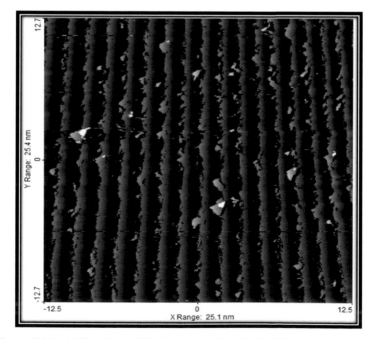

**Figure 9.34** A 25 × 25 nm STM image showing the individual SWCNTs in the monolayer film produced.

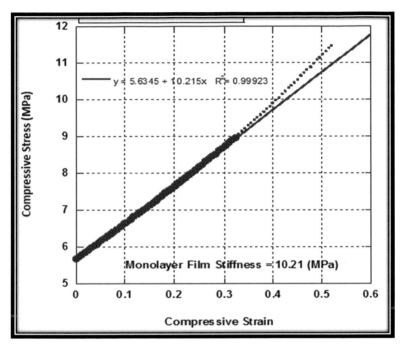

**Figure 9.35** Compressive stress–strain curve for the 2-D monolayer SWCNT based film. Note that compressive stress was applied normal to the tube axial direction.

Electrical conductivity measured for the SWCNT-based 2-D films was measured in both axial and transverse directions. The measured conductivity (see Figure 9.36) was $4.8 \times 10^4$ and $1 \times 10^4$ (S/cm) that is very close to the electric conductivity of individual isolated SWCNT which is known to be in the range of $10^4$ to $10^5$ (S/cm). While the electrical conductivity of such monolayer films is lower than that of bulk copper and silver metals (typically in the range of $6 \times 10^5$ S/cm) they are better electrical conductors than the Indium tin oxide (ITO) thin films ($10^3$ S/cm) and their transparency (98% in visible light range) is also higher than that of ITO films (typically 85% in visible light range). Hence, monolayer films based on highly oriented SWCNTs would be an excellent alternative for ITO in applications that require a transparent conductor.

It is also important to note that as such highly aligned SWCNT films were stacked on top of each other to create multi-layered films, their electrical conductivity dropped exponentially as the number of layers was increased as shown in Figure 9.37. The measured electrical conductivity of a 11 layers film was 0.07 S/cm which is almost an insulating film. This behavior in electrical conductivity is similar to the behavior in thermal conductivity as was discussed earlier.

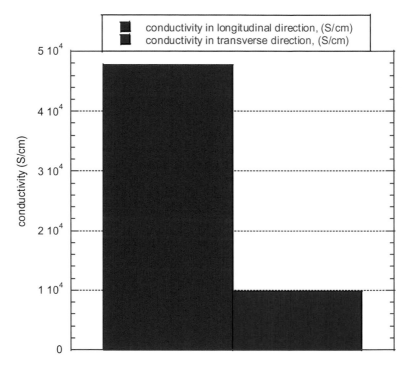

**Figure 9.36** Electrical conductivity for monolayer 2-D films of highly aligned SWCNTs in both axial and transverse directions.

## 9.13    SWCNTs Under Hydrostatic Pressure

The behavior of SWCNTs under hydrostatic pressure was experimentally investigated using spectroscopic techniques such as Raman spectroscopy. Raman results reported by different research groups show two distinctive characteristics; first, the peak positions of the radial breathing modes shift linearly to higher wavenumbers (blue shift) and their intensities decrease as the applied hydrostatic pressure increases, eventually disappearing (or becoming too weak to be detected) around an applied pressure of 2.5 GPa; secondly, the

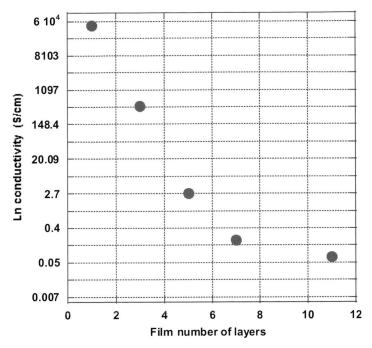

**Figure 9.37** Electrical conductivity of highly aligned SWCNT films as a function of number of film layers.

peak position of the tangential mode (especially the second order mode around 2660 cm$^{-1}$, the G'-band) is also blue shifted as the applied hydrostatic pressure increases, reaches a plateau at the same applied pressures at which the radial modes starts to disappear. Typical SWNT Raman spectral behavior for the peak position of both the radial breathing mode and the second order tangential mode (G'-band) are shown if Figure 9.38. RBM and G'-band peak position as function of applied hydrostatic pressure are shown in Figure 9.39.

The disappearance of the radial breathing mode and the independence of the second order peak position on applied pressure have both been interpreted previously as the result of lateral deformation of the nanotube as its cross-section changes from a circular into elliptical, hexagonal, and eventually flattens in what has been termed as ovalization, polygonization, or collapse of the nanotube. The same interpretation in terms of cross-section deformation has also been invoked to interpret X-ray and neutron diffraction results as well as a number of theoretical and simulation investigations.

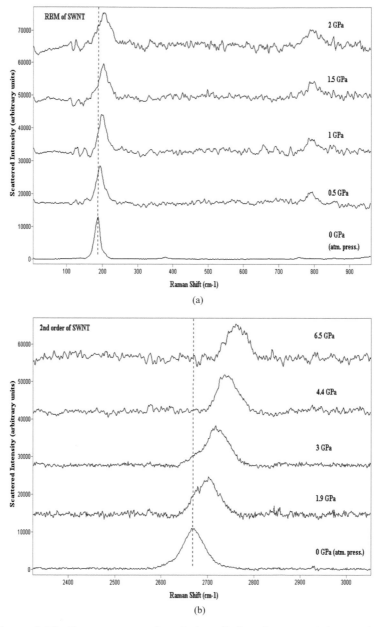

**Figure 9.38** Raman spectra for single-walled carbon nanotubes under successively increasing hydrostatic pressure. (a) Radial breathing mode and (b) G′-band. Reproduced with kind permission from Amer et al., *J. Chem. Phys.*, 2004, 121, 2752. Copyright American Chemical Society, 2004.

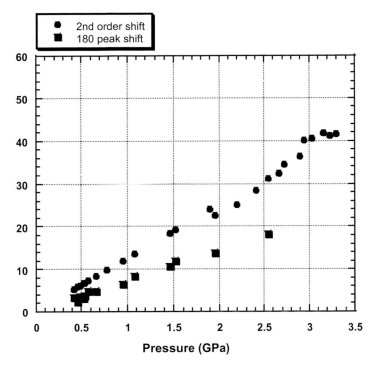

**Figure 9.39** Radial breathing mode and second order G′-band Raman peak positions as function of applied hydrostatic pressure in pure methanol. Reproduced with kind permission from Amer et al., *J. Chem. Phys.*, 2004, 121, 2752. Copyright American Chemical Society, 2004.

A different interpretation for the observed high pressure Raman behavior of SWCNT was introduced in 2004 relating the Raman behavior to molecular perturbation effects resulting from pressure transmission fluid molecular adsorption and interaction on the nanotube surface. The observed disappearance of the radial breathing mode according to this interpretation would be due to the possibility that adsorbed molecules on the nanotube surface takes the RBM out of resonance leading to its disappearance. This would definitely depend on the exact nature of the adsorbing molecule and the exact nature of its interaction with the nanotube. Figure 9.40 shows peak position of the radial breathing mode for three different SWCNTs [(10,10), (9,9), and (8,8)] dispersed in methanol, and water (using the known 1% SDS surfactant method), respectively, as a function of applied hydrostatic pressure. It is interesting to note that while the RBM mode for the (10,10) dispersed in methanol disappears around 3 GPa of applied pressure, it does not disappear up to 14 GPa of applied pressure for the nanotubes dispersed in water.

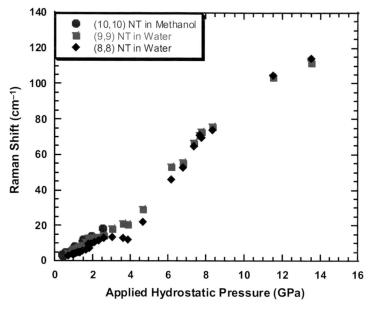

**Figure 9.40** RBM for three different SWNTs; (10,10) dispersed in methanol, (9,9) dispersed in water, and (8,8) dispersed in water, as a function of applied hydrostatic pressure. Note that the RBM for CNTs dispersed in water does not disappear up to 14 GPa of applied pressure. Figure courtesy of Prof. M. S. Amer, 2004.

It is important to note that on the molecular level, the molecules in the pressure transmission fluid intimately interact with the surface of the pressurized sample thereby transmitting the pressure to the bulk of the solid. For a typical high-pressure Raman investigation on macroscopic samples of characteristic length scale ($L$) in the range of millimeters or even micrometers (a bulk system), the interactions between the liquid phase, the pressure transmitting fluid (PTF) and the sample surface are not observable since the spectrum is completely dominated by the bulk phonons of the sample. In such a case, due to the relatively large interaction volume between the excitation laser and the sample (typically few cubic microns) none of the perturbation effects reflected in the surface phonons can be

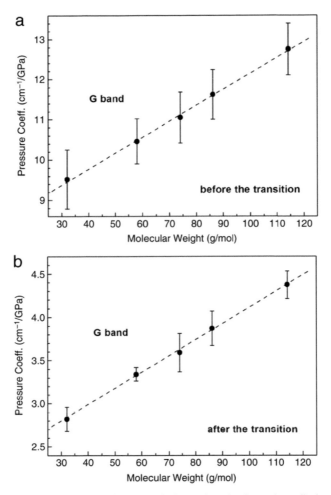

**Figure 9.41** Pressure coefficient of the G-band of single-walled carbon nanotube under hydrostatic pressure as a function of the molecular weight of different organic solvents used as pressure transmitting fluids, (a) before and (b) after the plateau appears. Reproduced with kind permission from Gao et al., *Solid State Communi.*, 2008, 147, 65. Copyright Elsevier, 2008.

detected. However, as the sample characteristic length scale ($L$) is greatly reduced into the nanometer range, surface effects become increasingly more important. Changes in surface phonons due to solvent interactions start to have a measurable effect on the observed spectral response of the sample. In particular, mesoscopic molecules like SWCNTs and spheroidal fullerenes, when dispersed as individual nanoclusters or molecules surrounded by solvent molecules (the PTF), have essentially no bulk phase as such but have only surface. In this case, the solvent interactions on the surface of the nanoparticles are expected to play a major role in Raman spectral changes. In fact, the Raman spectra in such a case should reflect, to a large extent, the perturbation effects due to solvent interactions on the surface of the fullerenes.

In addition, studies performed to investigate the effect of pressure transmitting fluid on the Raman peak shifts of SWCNTs using five different organic solvents, namely methanol, propanol, 1-butanol, hexane and octane showed a linear dependence of the pressure coefficient $(d\omega/dp)$ of the G-band on the molecular weight of the organic solvent used for transmitting the pressure. The pressure coefficient of the G-band (the slope of the linear dependence of the G-band position on applied hydrostatic pressure) was found to increase as the molecular weight of the organic solvent increased as shown in Figure 9.41. Such results further confirm that the shifts observed in Raman active bands of SWCNTs are related to perturbation fields (mechanical and chemical) rather than only mechanical as observed in bulk systems. It is also important to note that the shifts in the Raman bands are always correlated with changes in the system's electronic structure leading to changes in its electrical conductivity. Hence, it is possible to tune fullerenes' electrical conductivity by applying either mechanical and/or chemical perturbation fields.

## Problems

1. What are the different types of linear 1-D nanotubes?
2. What are the different classes of SWCNTs?
3. Express the circumference of a SWCNT in vector notation.
4. Define the chiral angle $(\theta)$ for a SWCNT.
5. Tube (10,5) and (5,10) are equivalent tubes. Explain the statement.
6. Determine the chiral angle for the following tubes:
   a. (8,8)    b. (8,0)    c. (8,3)    d. (10,10)
   e. (9,0)
7. Which of the SWCNT in Problem 6 is metallic and which is semiconducting?

8. One half of a $C_{70}$ molecule can correctly cap the ends of which SWCNT?
9. Determine the diameter of each SWCNT in Problem 6.
10. Which SWCNT can be capped by half a $C_{60}$ molecule?
11. Which tube configuration would result if a spherical molecule is bisected normal to one of its five-fold rotation axis?
12. Determine the number of carbon atoms in a unit cell of (18,0) nanotube.
13. Explain the difference between symmorphic and non-symmorphic space groups in SWCNT.
14. Determine the symmetry group the following SWCNTs:
    a. (8,8)    b. (8,0)    c. (8,3)    d. (10,10)
    e. (9,0)
15. In a MWCNT, determine the second and third layer tubes if the core tube is:
    a. (998)    b. (8,0)    c. (8,3)    d. (10,10)
    e. (9,0)
16. Carbon nanotube production methods are classified into two groups. Explain the two classes and different methods under each class.
17. Discuss the factors that are important in producing a high yield and good quality of carbon nanotubes.
18. Which catalyst was found to yield the maximum production in the arc-discharge method?
19. Explain The HiPco Process for SWCNT production.
20. Explain the purification process for carbon nanotubes.
21. How many Raman active modes in the following tubes: (10,10), (9,0), and (10,5)?
22. Determine the Raman position for the RBM in bundles of the following tubes: (18,0) and (10,10).
23. Why RBM does not appear in DWCNT and MWCNT?
24. Thermal conductivity of SWCNTs was found to depend on the structure of the tube. Explain this statement.
25. What is the predominant heat transfer mechanism within carbon nanotubes?
26. Is it possible to tune fullerene electrical conductivity? Explain your answer.

# Chapter 10

# Two-Dimensional Fullerenes, Planar Fullerene, or Graphene

## 10.1  Introduction

While the unique properties of a two-dimensional (2-D) form of graphitic carbon were predicted since 1947, the actual production of graphene samples in 2004, provided an unprecedented opportunity to investigate such unique 2-D material experimentally. The physics of single or few layers of graphene sheets was shown to be very unique and vastly different from the physics of bulk layered graphite. From thermodynamic viewpoint, the melting temperature of free-standing thin films should rapidly decrease with decreasing film thickness. Once film thickness reaches a limit of, typically, few tens of atomic distances, the film should become unstable, i.e., decompose or transform into particles (3-D structures) unless it is part of a 3-D system, i.e., grown on a substrate. However, and in spite of many arguments, more than half a century ago, stating strictly that 2-D crystals are thermodynamically unstable and cannot exist, single and few layers of graphene were prepared and reported in 2004. Graphene, in particular, dragged a lot of scientific attention due to its theoretically predicted unique physical properties such as high electron mobility at room temperature up to (250,000 $cm^2/Vs$), exceptional thermal conductivity (5000 W/mK), superior value of

*Gigantic Challenges, Nano Solutions: The Science and Engineering of Nanoscale Systems*
Maher S. Amer
Copyright © 2022 Jenny Stanford Publishing Pte. Ltd.
ISBN 978-981-4877-74-9 (Hardcover), 978-1-003-14704-6 (eBook)
www.jennystanford.com

elastic modulus in the range of 1 TPa, exceptional intrinsic strength close to 130 GPa, and calculated specific surface area ~2630 m²/g. The predicted superior electrical and thermal properties of graphene triggered a "graphene rush." In fact, publication statistics shows that roughly one paper per day were published with the term "graphene" in their title since 2004, reflecting the potential of such building block in future nanotechnology. Figure 10.1 shows the number of patents and publications on graphene since 2004 till 2018.

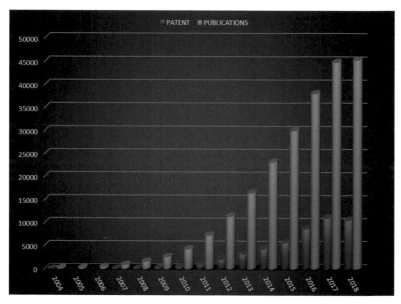

**Figure 10.1** Number of patents and publications on graphene since its first successful preparation in 2004. Courtesy of Web of Science, 2018.

## 10.2 The Structure

Graphene unit cell, with a lattice constant $a_1$= 2.46 Å, contains two independent lattice sites, A and B, as shown in Figure 10.2. With an average C–C bond length of $a$ = 1.42 Å, the unit cell area of graphene is 5.24 Å². In addition, realizing the six-fold rotation axis and the six mirror planes in the unit cell as shown in Figure 10.3, the graphene unit cell belongs to the P6mm 2-D space groups.

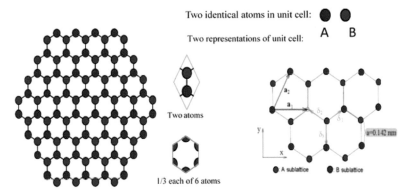

**Figure 10.2**  Schematic presentation of the graphene unit cell. 2 carbon atom per unit cell, unit cell area = 5.24 Å$^2$.

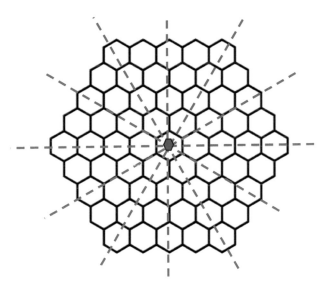

**Figure 10.3**  Symmetry elements in graphene.  One six-fold rotation axis $C_6$, and six mirror planes m.

Atomic structure of free standing single-layered graphene sheets was investigated using high resolution transmission electron microscopy (HR-TEM) (see Figure 10.4). While the hexagonal lattice symmetry of graphene was confirmed, HR-TEM images reveal the presence of triangular symmetries, which break the hexagonal lattice symmetry of the graphitic lattice.

**Figure 10.4**   HR-TEM image showing the structure of a standing graphene film. Courtesy of Prof. M. S. Amer, 2018.

In addition to structural imperfections in the film, structural corrugations of the graphene sheet deposited on substrates and conforming to the underlying substrate topography, emphasize the importance of substrate topography on the configuration, and hence the performance, of graphene sheets. Figure 10.5 shows a scanning tunneling microscopy (STM) image

Density functional theory (DFT) calculations were utilized to investigate the different possible types of point defects and their effect on electronic and optical performance of single graphene sheets. The three different types of defects investigated were a mono-vacancy (MV), a di-vacancy (DV), and a Stone–Wales (SW) defect. Figure 10.6 shows the optimized atomic structure of graphene with three different point defects: (a) an MV defect, (b) a DV defect, and (c) a SW defect. All three structures were found to be almost flat except for the MV one, where the unpaired atom has a large out-of-plane displacement of 0.49 Å.

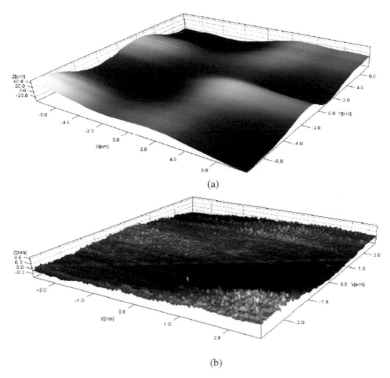

Figure 10.5 Scanning tunneling microscopy (STM) micrograph depicting a graphene sheet conforming to $SiO_2$ substrate topography. (a) A 16 × 16 nm scan, and (b) a 5 × 5 nm scan at the center of the 16 nm scan. Courtesy of Prof. M. S. Amer, 2017.

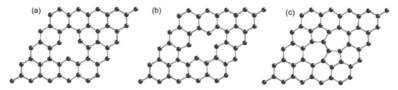

Figure 10.6 The optimized atomic structure of graphene with three different point defects: (a) a monovacancy (MV) defect, (b) a di-vacancy (DV) defect, and (c) a Stone–Wales (SW) defect. All three structures are almost flat except for the MV one, where the unpaired atom has a large out-of-plane displacement of 0.49 Å. Reproduced with kind permission from Popov et al., *Carbon*, 2009, 47, 2448. Copyright Elsevier, 2009.

The defects were found to modify the electronic structure and the phonons of graphene giving rise to new optical transitions and defect-related phonons. In addition, density functional theory and quantum Monte Carlo simulations of SW defects in graphene reported revealed that the structure of the SW defect in graphene is more complex than thus far appreciated. It was pointed out that rather than being a simple in-plane transformation of two carbon atoms, SW defects result in out-of-plane wavelike defect structures that extend over several nanometers.

## 10.3 Production of Graphene

Graphene has been produced by several methods. Some of these approaches including several lithographic, chemical and other synthetic procedures are known to produce microscopic samples of graphene. Macroscopic quantities of graphene were also produced using conventional as well as microwave plasma chemical vapor deposition. Figure 10.7 shows a schematic of the recently (2009) developed microwave plasma chemical vapor deposition instrument used to grow macroscopic quantities of graphene sheets.

Another unique approach to prepare graphene is by longitudinal unzipping of SWCNTs and MWCNTs using oxidation treatment. This approach is very well suited for research purposes since it enables the preparation of very well controlled samples for scientific investigations. The oxidation treatment involves suspending carbon nanotubes in concentrated sulfuric acid followed by treatment with 500 wt% $KMnO_4$ for 1 h at room temperature (22 °C). Figure 10.8a shows a schematic of the unzipping process. Figure 10.8b shows the oxidation reaction leading to the longitudinal unzipping of the nanotube into a graphene sheet. Figure 10.9 shows transmission electron micrographs depicting the transformation of multi-walled carbon nanotube into oxidized graphene sheets. Graphene sheets yield using this method is almost 100% and, hence, no further purification process is required. Table 10.1 summarizes the different production methods discussed above. Such methods are typically referred to as "bottom-up approaches."

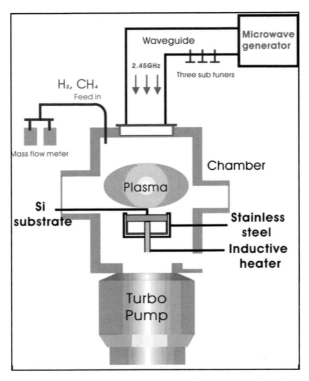

**Figure 10.7** Schematic of the micro-wave CVD system for the growth of graphene sheets. Reproduced with kind permission from Yuan et al., *Chem. Phys. Lett.*, 2009, 467, 361. Copyright Elsevier, 2009.

**Table 10.1** Different bottom-up approaches for graphene production

| Method | Product Nature | Advantage |
|---|---|---|
| Confined self-assembly | One layer at a time | Thickness control |
| CVD | Single layer/ multi-layers | High quality, large areas |
| Epitaxial growth on SiC | Multi-layers | High quality, large areas |
| Unzipping of carbon nanotube | Single layer/multi-layers | Size control |

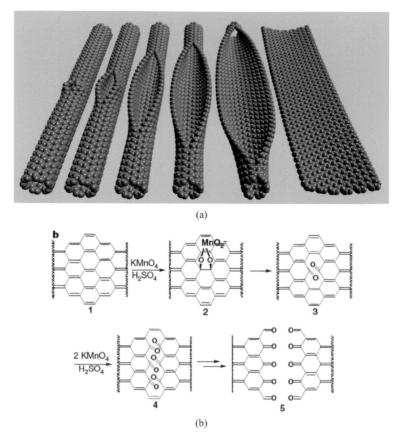

Figure 10.8 (a) Schematic of the longitudinal unzipping of single-walled carbon nanotube into a single graphene sheet. (b) The proposed chemical mechanism of nanotube unzipping into graphene sheets. Reproduced with kind permission from Kosynkin et al., *Nature*, 2009, 458, 872. Copyright Nature Publishing Group, 2009.

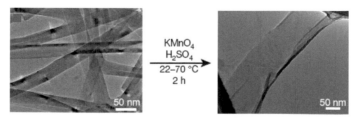

Figure 10.9 Transmission electron image depicting the transformation of MWCNTs (left) into oxidized graphene sheets (right). Reproduced with kind permission from Kosynkin et al., *Nature*, 2009, 458, 872. Copyright Nature Publishing Group, 2009.

**Table 10.2**  Exfoliation methods

| Exfoliation Method | Product Nature | Advantages |
| --- | --- | --- |
| Micro-mechanical | Single to few layers | Large size, defect-free GNFs |
| Liquid (solvent) Phase | Single + multi-layers | High quality graphene, cost effective |
| Electrochemical | Single + multi-layers | Single step functionalization |
| Shear mixing | Multi-layers | Large quantities, defect free |
| Super acid dissolution | Single layer | High-concentration solutions |

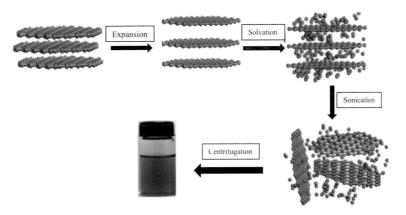

**Figure 10.10**  A schematic showing the steps of the liquid (solvent) exfoliation method.

Another approach to produce graphene is typically referred to as a "top-down approach." The process is based on exfoliating graphite into graphene sheets or graphene nanoflakes (GNFs). Several exfoliation methods have been applied successfully. This approach can be classified into oxidative and non-oxidative approaches. The oxidative approach starts with graphite oxide (GO) as a precursor. The graphite oxide has a larger interlayer spacing than that in graphite, hence, it is easier to exfoliate into a graphene oxide layers (GO) that can be reduced later into graphene. The non-oxidative approach, on the other hand, starts with graphite as a precursor. Table 10.2 summarizes the different exfoliation techniques and their advantages. Among all of the aforementioned exfoliation techniques, the liquid (or solvent) exfoliation technique has been considered the most promising method for bulk production of GNFs. Figure 10.9 shows a schematic illustration of the steps of liquid (solvent) exfoliation as an example for graphene exfoliation methods.

## 10.4    Raman Scattering of Graphene

Raman scattering, with its powerful ability to investigate vibration modes and perturbation effects has been heavily utilized in investigating graphene structural features and superior electrical and thermal performance. The Raman spectrum of graphene mainly exhibits two major Raman active modes: a G-band around 1580 cm$^{-1}$ and a G'-band (also referred to as 2-D-band in the literature) around 2700 cm$^{-1}$. Figure 10.11 shows Raman spectra of graphene and graphite for comparison.

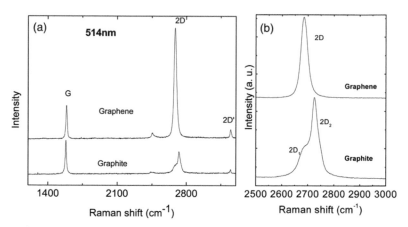

**Figure 10.11**    (a) Comparison of the Raman spectra of graphene and graphite measured using a 514.5 nm excitation laser. (b) Comparison of the 2-D peaks in graphene and graphite. Reproduced with kind permission from A. C. Ferrari, *Solid State Commun.*, 2007, 143, 47. Copyright Elsevier, 2007.

Raman scattering features of graphene were found to depend sharply on the number of graphene layers in the sample, thermal effects, and more interestingly, on the interaction between the graphene sample and its supporting substrate, and interaction with any dopants in the graphene sheet. Figure 10.12 shows the evolution of graphene Raman spectrum as the number of graphene layers increases. High-frequency 1st and 2nd order Raman spectra of graphene films supported on a $SiO_2$:Si substrate and that of HOPG. The data were collected using 514.5 nm radiation under ambient conditions. It is interesting to note that the shape and position of the Raman bands are sensitive to the number of layers ($n$) in the graphene film.

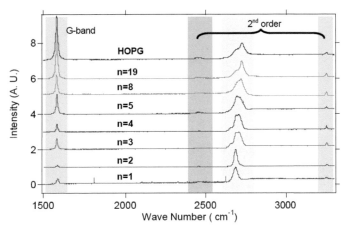

**Figure 10.12** High-frequency 1st and 2nd order Raman spectra of ($n$) graphene layer films supported on a $SiO_2$:Si substrate and that of HOPG. The data were collected using 514.5 nm radiation under ambient conditions. The spectra are scaled to produce an approximate match in intensity for the ~2700 cm$^{-1}$ band. Reproduced with kind permission from Gupta et al., *Nano Lett.*, 2006, 6, 2667. Copyright American Chemical Society, 2006.

Figure 10.13 shows the evolution of the G-band peak intensity and position as a function of film's number of graphene layers ($n$). It was reported that the G-band position shifts linearly to lower wavenumbers as the number of graphene layers increases. The observed dependence of the intensity of the G-band on the number of graphene layers was recently utilized in generating Raman images of graphene sheets distinguishing the number of layers of the graphene sheet based on the integrated intensity of its G-band. Figure 10.14 shows optical image of graphene with 1, 2, and 3 layers (upper), and Raman image plotted based on intensity of the G band (lower).

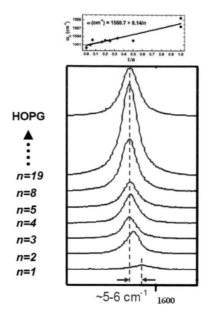

**Figure 10.13** Evolution of the G-line in graphene films with different number of layers. The data were collected using 514.5 nm radiation under ambient conditions. Reproduced with kind permission from Gupta et al., *Nano Lett.*, 2006, 6, 2667. Copyright American Chemical Society, 2006.

It is important to emphasize that while Raman intensity of the G-band (with its reported linear dependence on the number of graphene layers up to 10 layers) was suggested as an excellent, precise, and quick method to determine the number of layers in graphene, extreme caution should be taken while applying such a method. Based on our previous and following discussions, it should be clear that unaccounted for perturbation effects could easily render such method misleading. In addition, it is important to note that the exact interaction between successive graphene layers (definitely including the exact orientation of the graphene layers) will affect the collective Raman spectrum of the graphene sheet. Investigations of single, double, and two (folded) graphene layers showed that the Raman spectra of the folded two layers graphene is different from a double-layer graphene as shown in Figure 10.15.

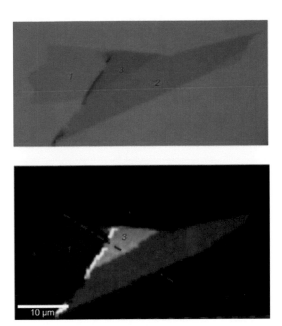

**Figure 10.14** Optical image of graphene with 1, 2, and 3 layers (upper), and Raman image plotted by intensity of the G band (lower). Reproduced with kind permission from Ni et al., *Nano Res.*, 2008, 1, 273. Copyright Springer, 2008.

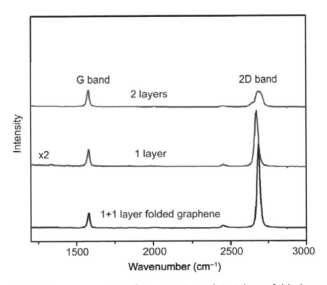

**Figure 10.15** Raman spectra of DLG, SLG, and 1+1 layer folded graphene. The spectra are normalized to have similar G band. Reproduced with kind permission from Z. Ni et al., *Phys. Rev. B*, 2008, 77, 235403. Copyright American Physical Society, 2008.

Disorder (or double resonance) induced D-band around 1360 cm$^{-1}$ was also observed in graphene Raman characteristic spectra. As shown in Figure 10.16, the D-band intensity changes with the number of graphene layers in the film. Strains induced in the graphene film due to substrate roughness (1–2 nm) and film edges are thought to be the reason for the observed D-band.

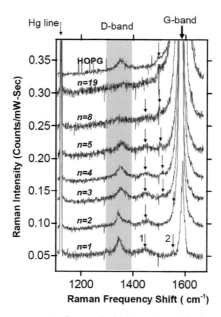

**Figure 10.16** Raman spectra (expanded intensity scale) of supported graphene films with (*n*) layers and HOPG showing the D-band. The data were collected using 514.5 nm radiation under ambient conditions. The arrows in the figure identify weak scattering features. Mercury (Hg) line was used for calibration purposes. Reproduced with kind permission from Gupta et al., *Nano Lett.*, 2006, 6, 2667. Copyright American Chemical Society, 2006.

The perturbation effect of the supporting substrate on Raman active modes of graphene was also investigated and reported. Room-temperature peak positions of Raman active modes (G-band and G′-band) from single-layer graphene interacting with different substrates showed significant shifts (up to 11 cm$^{-1}$) depending on the substrate. Figure 10.17 shows Raman spectra of single-layer graphene interacting with different substrates. The exact positions and FWHM of the G- and the G′-bands are shown in Table 10.3.

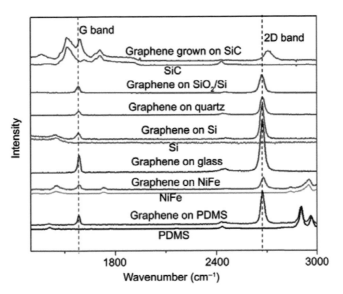

**Figure 10.17** Raman spectra of single-layer graphene deposited on different substrates. Reproduced with kind permission from Wang et al., *J. Phys. Chem. C,* 2008, 112, 10637. Copyright American Chemical Society, 2008.

**Table 10.3** Position and FWHM of Raman active modes in graphene monolayer on various substrates

| Substrate | G-band Position (cm⁻¹) | G-band FWHM (cm⁻¹) | G′-band Position (cm⁻¹) | G′-Band FWHM (cm⁻¹) |
|---|---|---|---|---|
| SiO₂/Si | 1580, 1580.5 | 15 | 2710.5 | 59.0 |
| GaAs | 1580 | 15 | | |
| Sapphire | ᵃ1575 | 20 | | |
| Glass | ᵃ1580*, ᵇ1582.5 | 35,16.8 | 2672.8 | 30.8 |
| Si | ᵇ1580 | 16 | 2672 | 28.3 |
| Quartz | ᵇ1581.9 | 15.6 | 2674.6 | 29.0 |
| NiFe | ᵇ1582.5 | 288.9 | 2678.6 | 30.8 |
| PDMS | ᵇ1581.6 | 15.6 | 2673.6 | 27 |

*G-band split on this substrate, the value represents the middle frequency.

Another important perturbation effect to consider is the effect of applied electric field (applied voltage) across graphene films. Graphene is very promising for electronic device applications and has been shown to demonstrate unique electronic performance best demonstrated by unique properties of charge carriers in graphene such as particle–hole symmetry of Dirac fermions. The coupling of long wavelength optical phonons (the G-band) with Dirac fermions was found to display remarkable changes in frequency and line-width that are tunable by the electron field effects. Applying voltage across a single-layer graphene sheet was found to significantly shift the G-band position of the graphene. Shift versus applied voltage enable the determination of the Dirac point of the graphene sheet. Figure 10.18 shows Raman spectra of single-layer graphene sheets depicting the G-band at different applied gate voltages at temperatures of 10 K (left) and 300 K (right).

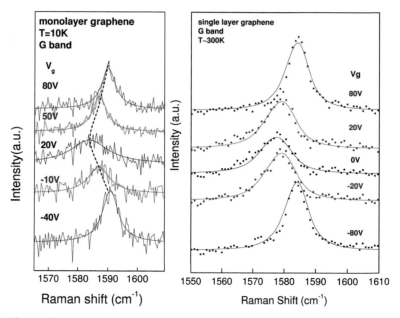

**Figure 10.18** Raman spectra of single-layer graphene sheets depicting the G-band at different applied gate voltages ($V_g$) at temperatures of 10 K (left), and 300 K (right). Reproduced with kind permission from Yan et al., *Solid State Commun.*, 2007, 143, 39. Copyright Elsevier, 2007.

Density functional theory DFT calculations investigating the impact of different types of point and topological defects, namely MV,

DV, and SW defects (see Figure 10.6), on Raman spectra of graphene single sheets showed that such defects would induce new phonons resulting in new Raman active mode to be observed in the graphene Raman spectrum. Based on the calculated Raman spectra, Raman lines that can serve as signatures of the specific type of investigated defects can be determined. In addition, it was predicted (based on the calculations) that the presence of defects would enhance the intensity of the G-band of graphene up to one order of magnitude compared to defect-free graphene. Figure 10.19 shows the calculated Raman spectra of graphene with different point defects for five laser photon energies EL: (a) MV, (b) DV, and (c) SW. Only the frequency region with more prominent Raman features is shown, and some of the spectra are scaled for better presentation. It is important to note that the calculation results presented in Figure 10.19 are for a free standing single graphene sheet in pure vacuum. The positions, intensities and resonance ranges for the predicted Raman modes would definitely be affected by substrate interaction, environment, and other experimentally unavoidable conditions.

## 10.5   2-D Films Based on Graphene Flakes

While many productions techniques were developed to produce graphene, as we discussed earlier, no technique was available to utilize the industrial production level available. The full realization of the unique properties of graphene required producing graphene films on a scale suitable for engineering applications. Recently, the Langmuir–Blodgett (LB) film production technique was used to produce large (in the range of tens of millimeters), monolayer graphene films using commercially available graphene nanoflakes (GNFs). The films could be deposited on a variety of substrates (such as metals, ceramics, polymers, etc.), and more importantly enabled better investigation, hence, understanding of the graphene film behavior and properties. Figure 10.20 depicts the LB isotherm for a moo-layer graphene films produced using GNFs. The isotherm shows that at zero-surface pressure, the area per unit cell of the film is 5.1 Å$^2$ which is very close to the theoretical value of graphene unit cell area of 5.24 Å$^2$ as was discussed earlier. Figure 10.21 depicts a STM current-based image of the film. Individual GNFs can be clearly seen in the image.

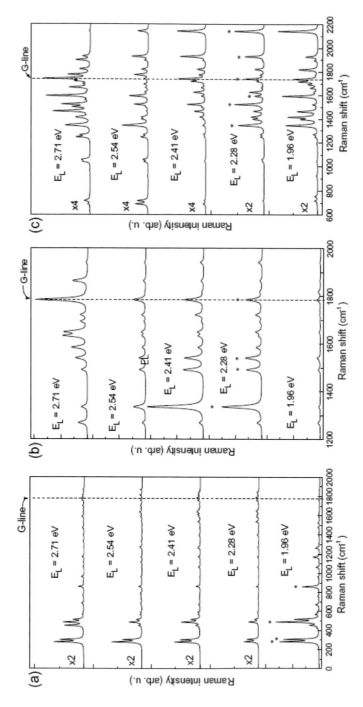

**Figure 10.19** The calculated Raman spectra of graphene with different point defects for five laser photon energies EL: (a) MV, (b) DV, and (c) SW. Only the frequency region with more prominent Raman features is shown. Some of the spectra are scaled for better presentation. The most intense Raman lines are denoted by asterisks. Reproduced with kind permission from Popov et al., *Carbon*, 2009, *47*, 2448. Copyright Elsevier, 2009.

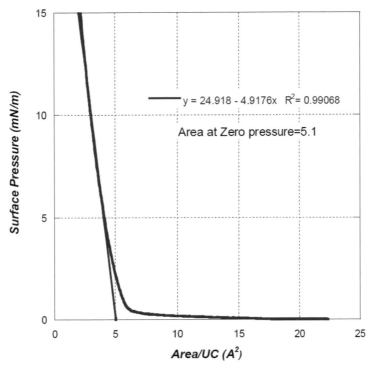

**Figure 10.20** LB isotherm for a monolayer graphene film made using GNFs (Image created by Prof. M. S. Amer, 2020).

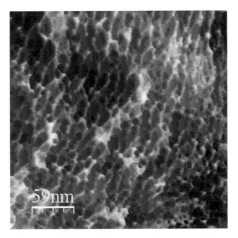

**Figure 10.21** STM current-based image of a monolayer graphene film produced using GNFs (Image created by Prof. M. S. Amer, 2020).

## 10.5.1 Mechanical Properties

Figure 10.22 depicts the stress strain curve for monolayer graphene film under compression. The non-linear behavior is clear from the curve. The stress–strain equation of the monolayer film fits a second order polynomial of the form

$$\sigma = 41.87\varepsilon - 44.8\varepsilon^2$$

Such strain stiffening behavior fits the predictions of simulation theoretical studies. The stiffness of the monolayer film ($E = d\sigma/d\varepsilon$) was found to increase linearly from 42 MPa to over 80 MPa as the compressive strain reaches the value of 50% as shown in Figure 10.23. It is very important to note that such non-linear elastic behavior under compression was observed in neither $C_{60}$-based films nor SWCNT-based films. The stiffness of the GNFs-based film is also high compared to the stiffness of the other two types of films.

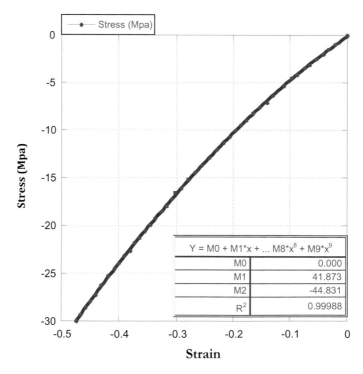

**Figure 10.22** Compressive stress–strain curve of a monolayer graphene film depicting non-linear behavior (Image created by Prof. M. S. Amer, 2020).

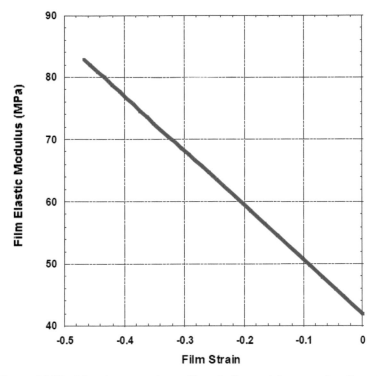

**Figure 10.23** Monolayer graphene film elastic modulus as a function of applied compressive strain (Image created by Prof. M. S. Amer, 2020).

## 10.5.2 Optical Properties

Monolayer graphene films deposited on glass substrate showed a full transparency in the range 400 nm and 1100 nm wavelength as shown in Figure 10.24. It is important to note that free standing graphene films supported on steel showed 83% transparency within the same range. Such difference in the optical absorbance of the films was attributed to suppression of out-of-plane vibration mode and the substrate interaction effects.

## 10.5.3 Electrical Conductivity

Electrical conductivity of the monolayer GNFs-based films was measured in two perpendicular directions and was found the same indicating an isotropic film behavior as shown in Figure 10.25. The

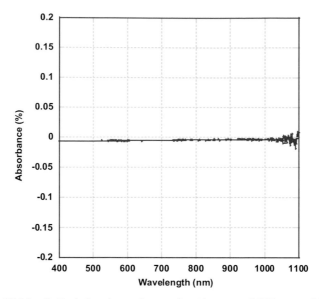

**Figure 10.24** Optical absorbance in wavelength range of 400 nm to 1100 nm for a monolayer, 4 layers, and 6 layers graphene films (Image created by Prof. M. S. Amer, 2020).

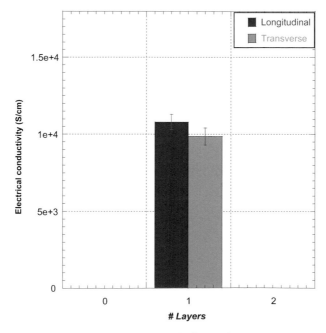

**Figure 10.25** Electrical conductivity of a (LB) monolayer graphene film in two perpendicular directions (Image created by Prof. M. S. Amer, 2020).

measured electrical conductivity was $1.08 \; 10^6$ S/m. This value is the same value measured for monolayer continuous graphene films produced using CVD production method. It is also the same for the most used transparent conductor ceramic ITO. It is important, however, to note that such a value for electrical conductivity is two orders of magnitude lower that the theoretical limit for the electrical conductivity of graphene estimated at $1 \times 10^8$ S/m.

## 10.5.4   Graphene to Graphite

Giving the superior properties of graphene and the not so unique properties of graphite, several investigations addressed the question "After how many layers graphene will become graphite?" Experimental investigations based upon spectroscopy, thermal and electrical conductivity as well as theoretical calculations results put the limit between 18 and 20 layers. Figure 10.26 depicts the absorbance of a monolayer graphene film, a 4-layer film, and a 6-layer film within the range 400 nm to 1100 nm. It is clear that all films are totally transparent within the investigated range of wavelengths.

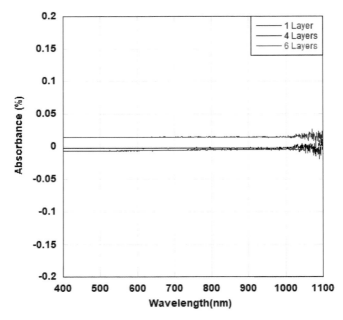

**Figure 10.26**   Optical absorbance in wavelength range of 400 nm to 1100 nm for a monolayer, 4 layers, and 6 layers graphene films (Image created by Prof. M. S. Amer, 2020).

Electrical conductivity measurements showed an exponential decrease in the film's conductivity as the number of graphene layers increases. The film electric conductivity matches the in-plane graphite conductivity at 18 layers as shown in Figure 10.27. Density functional theory (DFT) calculations of the films energy gap showed the same trend indicating that the drop in electrical conductivity is a direct result of the increase in the film's energy gap.

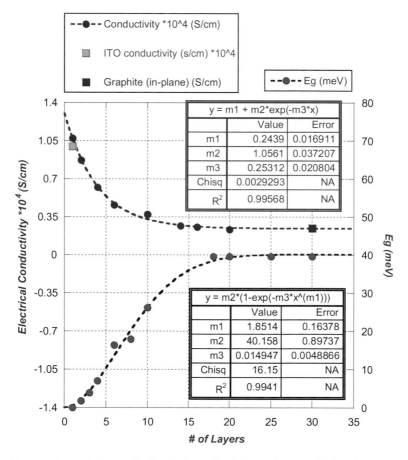

**Figure 10.27** Measured electrical conductivity and DFT calculated energy gap for graphene films as a function of graphene number of layers in the film. Reproduced with kind permission from Amer et al., *Philos. Mag.*, 2020, 100(19), 2491–2502. Copyright Taylor & Francis, 2020.

## Problems

1. Does the melting point of a free-standing thin film increases or decreases as film thickness decreases?
2. State some of the superior properties of graphene.
3. Calculate the area per unit cell for graphene.
4. What are the different types of point imperfections possible in graphene?
5. State the different techniques used to produce graphene.
6. What is the most important Raman feature of graphene?
7. Would the Raman features of graphene change as graphene is deposited on different substrate? Explain your answer.
8. Would point defects affect the Raman spectrum of graphene? Explain your answer.
9. Which elastic behavior does graphene exhibit? Explain your answer.
10. It was found that optical transparency of free standing graphene is much lower than that of films deposited on substrates. Explain such behavior.
11. After how many layers does a multi-layered graphene film exhibit electrical properties similar to graphite?

# Chapter 11

# Overview, Potentials, Challenges, and Ethical Consideration

## 11.1 Overview

In this book we started with the *supposedly* simple question: What is nanotechnology? The answer we strived to provide, with nearly fifty thousand words, implies that the question was, indeed, not so simple. In fact, here, we tried to capture the essence of what most plausibly would shape the future of humankind. We defined nanotechnology, in non-scientific terms, as the technology that would enable us to bridge what is natural and what is human-made. We have pointed out that nanostructured systems perform in a manner that is neither customary to us nor can be explained within the frame of knowledge we developed for *bulk* systems.

We showed that within the frame of our knowledge the best tools—thermodynamic and statistical mechanics—we developed to describe the behavior of materials are incapable once the system approaches certain length-scales. We showed that once the system approaches such length-scales, fundamental thermodynamic quantities such as temperature, pressure, free energy, chemical potential, and surface tension become indefinable, and therefore new physics, chemistry, and materials science are needed to enable describing the new system behavior. These combined new fields,

*Gigantic Challenges, Nano Solutions: The Science and Engineering of Nanoscale Systems*
Maher S. Amer
Copyright © 2022 Jenny Stanford Publishing Pte. Ltd.
ISBN 978-981-4877-74-9 (Hardcover), 978-1-003-14704-6 (eBook)
www.jennystanford.com

in fact, represent a new multidisciplinary field—nanoscience and technology—that ought to be developed and understood. We also showed that while critical or characteristic length-scales can be on the order of few millions of light years, the most interesting length-scale for scientist and engineers on planet earth is the nanometer. The nanometer gains its magic power from the fact that it is the distance over which atoms and molecules on our planet can correlate together via a number of energetic interactions forcing the system to enter the nanodomain. Once in the nanodomain, a material system's behavior becomes dominated by quantum effects, thermal fluctuations, and entropic effects. Thus, the system starts to behave unconventionally and exhibits *size and shape dependent* chemical and physical properties.

We showed how a slight (10 extra atoms) change in molecular size between $C_{60}$ and $C_{70}$ would alter its interaction with the same solvent. We also showed how changing the shape of a molecule from spherical ($C_{60}$) into cylindrical (single-walled nanotube) would turn it insoluble in the same solvent. In addition, we showed how optical and electrical properties of a nanosystem would significantly depend on the system's size and shape.

We also showed how a nanosystem would, amazingly, interact with its environment. It was shown that as much as the fullerene properties are altered by perturbation effects imposed by a liquid environment around it, it also affects the structure, and hence the properties of such liquid. Future investigations will definitely reveal more amazing phenomena of the nanodomain, or more amazing nanophenomena.

## 11.2  Potential

It is very well agreed upon that nanoscale science and engineering will advance our understanding of nature and materials systems. Such advancement will lead to significant changes in manufacturing, the economy, healthcare, and many other aspects of our standard and way of living. Examples include the following:

- *Advanced Materials:* As we have discussed earlier, nanostructured materials exhibit higher performance than traditional bulk materials created by traditional chemistry. Hence, nanostructured materials are predicted to advance many technologies.

- *Manufacturing*: Once nanoscale length is adapted, it is expected to become a highly efficient length scale for more precise, highly efficient manufacturing technology.
- *Electronics*: Nanotechnology has already impacted the electronic industry. Better processors and more advanced computational capabilities will definitely cause advances in many other fields. is projected to yield annual production
- *Pharmaceuticals*: It is predicted that about half of all pharmaceutical production will be dependent on nanoscale science and engineering affecting leading to more effective medications with less side effect and less cost.
- *Healthcare*: Realizing that living systems are controlled by molecular behavior at the nanoscale, nanoscale science and engineering with its ability to create improved pharmaceuticals, advanced medical characterization tools, and advanced implants, will result in longer lifespan, and extended human capabilities.
- *Energy:* Efforts utilizing nanoscale science and engineering have shown a great potential to significantly impact energy efficiency, storage, and production. For example, nanoscale-based developments showed increase in solar cell the efficiency, high efficiency fuel cells, and more effective catalysts.
- *Fresh Water Resources:* Depletion of the global fresh water resources is daunting. The United Nations predicts that by the year 2025, 32% of the world's population in 48 countries will lack the needed resources of fresh water. Nanoscale-based devices for water desalinization are thought to be at least 10 times and 100 times more energy efficient than reverse osmosis and distillation-based devices, respectively.

To conclude regarding the potentials, nanophenomena represent a new domain of human knowledge. However, it is nothing that we can claim inventing. Nature for over 3 billion years has been utilizing nanophenomena and nanomanufacturing in creating the most versatile, and efficient system that is capable of sustaining itself—*life*—as we know it, and as we still trying to reveal it secrets. In his remarkable book *What is Life?* Erwin Schrödinger pointed out that life[1] is based upon "aperiodic crystals" while non-living material systems, as we know them, are based on "periodic crystals." Within the frame of bulk system laws, it is not simple to comprehend the formation mechanism or the stability of aperiodic crystals. This is why we refer to it as "the miracle of life." We currently refer to aperiodicity in a material system as "faults" or "defects." However, with our current crumb of knowledge in nanoscience, we are in a position allowing us to appreciate nature's marvelous aperiodic crystal-based or nanodomain-based designs. In fact, we are currently in a position enabling us to add to Schrödinger's viewpoint by stating that "life is based on nanosystems."

Once we moot many of the miracles of life, we cannot avoid realizing directed-assembly[2] of nanosystems reacting significantly to minute changes in their environment. More importantly and amazingly, we realize that such systems possess a unique ability of self-healing within certain limits. Examples are numerous. The ability of plant roots to grow in the correct direction under the effect of gravitational field, and the ability of sun flowers to control their stem stiffness to enable them to follow the sun under photo-effects are among the well-known examples.

Nowadays, we started to appreciate the potential of directed assembly on nanosystems. We have shown how the shape of solvent molecules would direct fullerene molecules to assemble themselves into zero-, one-, or two-dimensional shapes. We also started to investigate self-healing systems. While we are still far from being capable of assembling systems resembling those done by nature, it will not be exaggerating to say that future generations of scientists and engineers will be well positioned for that task.

---

[1]Here, life is used to denote living systems and not the ideologies and principles shaping human behavior.

[2]Self-assembly is the more frequently used term in the literature. Directed assembly is used here to emphasize that the atoms and molecules would assemble themselves under the effect of external fields in a way that is directed by the exact nature of the interaction with the field. Both terms are accepted in the scientific community.

## 11.3   Challenges

With so many potentials for nanoscale science and engineering, it will definitely impact the future trajectory and behavior of human society. Some of the challenges has been observed but more are expected to surface.

- *Large scale production and economic challenges*: Based on experience with earlier technologies, the common model is that new products will be initially more costly than existing technologies but offering better performance. In addition, serious investments in new production facilities are required. Because of that nanotechnology-based products are introduced earlier in markets where performance essential while cost is a secondary consideration. This includes medical applications, electronics, and defense markets. In a later stage, the experience gained in performance-oriented markets will reduce the cost and prepare the technology for deployment into cost-oriented markets. As a given technology matures, its cost may decline, leading to greater penetration of the market even where neither performance nor cost is essential such as disposable products. The domination of nanoscale-based technologies will also necessitate the development of a network of complementary technologies and totally new industries may have to be developed. This will definitely impact our workforce and education system since new trends of training and education will be required.

- *Impact on Society*: Economists and sociologists have studied the impact of previous scientific and technological advances on society. However, the impact of nanoscale technology, with its ability to bridge natural products and human-made products, on society is very complex and not yet fully understood. While scientific advances do not change society directly; they set the stage for such change. The change happens through the merging of old and new technologies directed by evolving economic and social aspects. Nanoscale technology potentials are so diverse that it will take decades to impact the socio-economic system. It is important to note that economic and social acceptance will ultimately determine the rate and transition time at which nanoscale technology will impact the socio-economic system.

Also, new technologies typically have an unintended impact on societies. Some of such unintended impacts or changes can be beneficial or otherwise. For example, the World Wide Web was originally developed for data communications between US department of defense and its contractors. None of the developer intended for the ever-increasing dependence of society on internet. Also, once the world became so much connected through the internet new and unprecedented economies and social behavior started to develop. Another example involves the high-quality medical care that nanoscale technology would enable and its impact on lifespan. Such increase in lifespan will require changes in pensions and an increase in the retirement age.

Also, a wide income and wealth gaps are expected to develop between nanoscale technology-based economies and other economies for people and countries that cannot afford to benefit from the new technology. Such gaps would lead to local or global unrest.

- *Ethical Issues*: Considering the societal impact discussed above, it become ethically important to address issues involving social justice, human rights, and professional ethics. For example, the predicted impact of the new technology on workforce and the need for human resources will benefit some and harm others. Also, nanoscale-based technology will be, at least during first stages of its applications, affordable to the wealthy part of society. As we mentioned before, this affordability issue will widen the gap in wealth and cause restructuring of society with many unfair and unethical consequences.

Most importantly, personnel involved in the development of the powerful nanoscale-based technologies and products must be well trained in identifying ethical issues relevant to their work. It will be crucial to incorporate ethics effectively into the science, engineering, and technical curricula.

After all, regardless of any other considerations, it is important to emphasize that the wealth of information that has been generated has significantly enhanced our understanding of how to develop nanostructured systems that, if utilized wisely, would definitely shape our future.

## Problems

1. State the potentials of nanoscale-based technology.
2. State the challenges for nanoscale-based technology.
3. What is the impact of nanoscale-based technology on society?
4. What are the ethical issues to consider once nanoscale-based technology is utilized?

# Index